The Invisible Classroom

The Norton Series on the Social Neuroscience of Education
Louis J. Cozolino, Mary Helen Immordino-Yang, Series Editors

The field of education is searching for new paradigms that incorporate our latest discoveries about the biological underpinnings of processes related to teaching and learning. Yet what is left out of these discussions is a focus on the social nature of human neurobiology, the interactive context of learning, and the quality of student-teacher relationships. The "interpersonal neurobiology of education" is a fresh perspective that will help teachers and administrators better understand how the relationships among educators, students, and the social environments they create within classrooms and schools promote brain development, support psychological health, and enhance emotional intelligence.

The Norton Series on the Social Neuroscience of Education publishes cutting-edge books that provide interdisciplinary explorations into the complex connections between brain and mind, social relationships and attachment, and meaningful learning. Drawing on evidence from research in education, affective and social neuroscience, complex systems, anthropology, and psychology, these books offer educators and administrators an accessible synthesis and application of scientific findings previously unavailable to those in the field. A seamless integration of up-to-date science with the art of teaching, the books in the series present theory and practical classroom application based on solid science, human compassion, cultural awareness, and respect for each student and teacher.

Norton Books in Education

The Invisible Classroom

Relationships, Neuroscience & Mindfulness in School

Kirke Olson

Foreword by Louis Cozolino

W. W Norton & Company
New York • London

Note to Readers: Models and/or techniques described in this volume are illustrative or are included for general informational purposes only; neither the publisher nor the author(s) can guarantee the efficacy or appropriateness of any particular recommendation in every circumstance.

Printed in the United States of America
First Edition

For information about special discounts for bulk purchases, please contact W. W. Norton Special Sales at specialsales@wwnorton.com or 800-233-4830

Manufacturing by LSC Harrisonburg
Book design by Paradigm Graphics
Production manager: Leeann Graham

Library of Congress Cataloging-in-Publication Data
ISBN 978-0-393-70757-1 (pbk.)

W. W. Norton & Company, Inc., 500 Fifth Avenue, New York, NY 10110
www.wwnorton.com
W. W. Norton & Company Ltd.,
15 Carlisle Street, London W1D 3BS

3 4 5 6 7 8 9 0

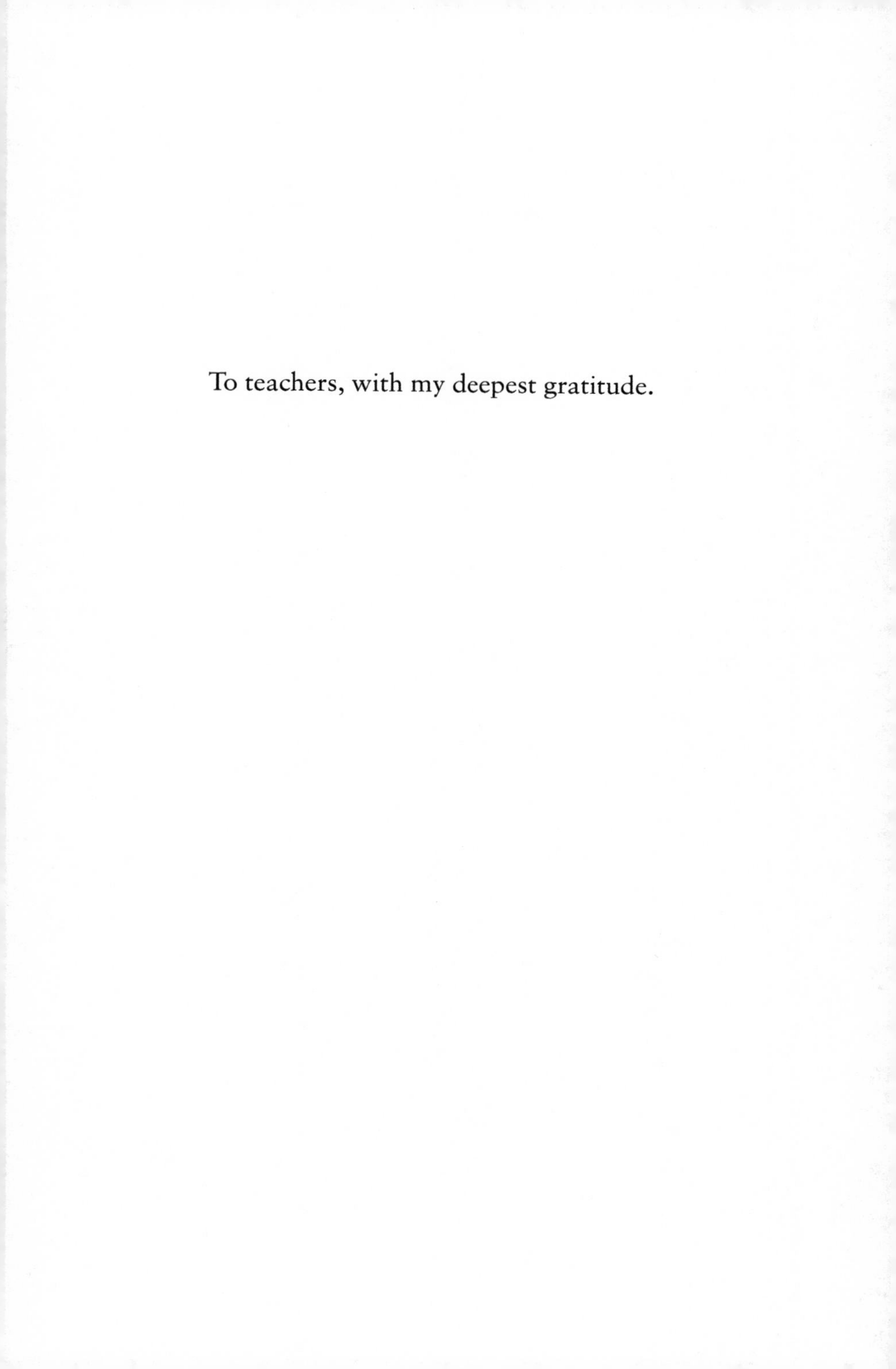

To teachers, with my deepest gratitude.

Contents

Acknowledgments

I HAVE BEEN BLESSED WITH THE SUPPORT OF TWO AMAZING AND powerful women, Sher Kamman and Bonnie Badenoch. I am proud to be the husband of the former—a gifted psychologist, workshop facilitator, and landscape photographer who demonstrated unending patience as it slowly dawned on me that the skill set for doing the work described here is quite different from the skill set required to write about it. I lovingly blame the latter woman—an accomplished author, therapist, and teacher—for getting me into this in the first place. She edited my doggerel, making it appear that I know more than I do. These two women worked as a team to pull me out of my dark night of the soul, which I knew to be a familiar experience for some authors, but one I naively believed I could escape. Deborah Malmud was a third member of this powerful team as she shone the light of reality and led this first-time author through the maze of the publication process and, with her team including Ben Yarling and Katie Moyer, improved my prose in the service of you, the reader.

This book simply would not exist without the cooperation of innumerable teachers and administrators in the public and private schools of New Hampshire. Your numbers are legion, and although your names and faces pass through my mind as I write this, limited space makes it impossible to list you all. You and I know who you are, and I wholeheartedly thank you. Most recently I must thank Dave Parker and the inspired and patient staff, students, and parents of Parker Academy in Concord, New Hampshire. They were willing to try out the ideas I suggested, showing me which ones worked and enthusiastically informing me about those that did not.

I owe a debt of gratitude to the many thought leaders who have inspired and taught me, especially Dan Siegel, Marty Seligman, the late poet John O'Donohue, Louis Cozolino, Allan Schore, Stephen Porges, Barbara Fredrickson,and the late Chris Peterson. The list could go on, but certainly would be incomplete without the members and board of GAINS (Global Association of Interpersonal Neurobiology). GAINS is a dedicated group and a deep resource for anyone interested in interpersonal neurobiology.

My forty-plus years of work in public and private schools and my own education have put me in contact with many gifted teachers. The lessons you taught are the back story of the *Invisible Classroom*.

Preface

I WILL NEVER FORGET MY HARSH INTRODUCTION TO THE INVISIBLE classroom—those hidden human connections among us and the microscopic neural connections inside of us that are critical for teaching and learning. It was my first day of high school teaching, which was also the day that began the development of my deep, abiding respect for teachers. I had just been honorably discharged from the navy after completing a transition program that trained me as a preschool teacher for children with serious emotional needs. I loved every minute of preschool teaching, so I found what I naively thought was a similar job replacing a charismatic and beloved teacher in a high school for students with serious emotional difficulties. Before that first day, I carefully planned my science, math, and physical education lessons for the twenty-seven teenage boys who were in my classroom instead of in jail or the hospital. Midway through the first morning, a boy much larger than me stood up, made a fist, stuck up his thumb, and slowly pulled it across his throat while staring at me in an unmistakable gesture of cutting my throat. My heart stopped, and the mood in the classroom shifted. I had worked hard to plan lessons that picked up where the popular previous teacher left off. The lessons took into account the wildly varied reading and math levels of the boys and appeared (to me) to be interesting. That gesture was my first introduction to the power of the invisible classroom—a collection of continuously active neurological and human connections that have an immense effect on learning but little to do with the actual content of the lesson. These undercurrents are always present, sometimes intuited, but rarely perceived and addressed directly.

In the tumultuous months following that first day, the boys repeatedly taught me the power of these invisible forces while I tried—sometimes successfully, sometimes not—to teach the course content. During one science class I was trying to teach about the plants growing directly outside of the classroom window, hoping to spark the students' enthusiasm about their natural environment, I turned to face the window to point out the distinctive bark of a birch tree. I heard giggling behind me. I turned to see each boy one by one feign sleep and slowly lower their heads down to their desks. Some even pretended to snore. My lack of knowledge at that time led me to make things worse by becoming angry and punishing the entire class. In hindsight, I realize I was not teaching a random collection of separate individuals; I was teaching a group of boys who were connected with each other and to some extent united against me, their new teacher. My punishment further strengthened the connections among the boys and hardened their resolve against me. I faced a long and tiring semester.

Since then I have learned from decades of experience with students, teachers, administrators, and parents while holding various positions in rural and small city schools, ultimately returning to school myself to take on the roles of a nationally certified school psychologist and state-licensed clinical psychologist. In these positions, teachers and administrators gave me the opportunity to have a close view of the workings of their classrooms and schools from the enviable position of being simultaneously an insider and an outsider. Because I was an insider and trusted member of the school staff, teachers revealed their struggles with a wide range of classroom problems, asking me to observe, solve problems, and sometimes teach their classes. They gave me an intimate view of the classroom as we collaborated to solve student, classroom management, and even personal problems. The experience has given me a deep understanding of the intense emotional, cognitive, social, and physical demands of day-to-day life in a classroom that leave teachers precious little time to step back and reflect. I am an outsider to the classroom because as a psychologist I was trained differently and don't spend every minute of the day immersed in teaching. This allows me to have a different perspective as well as the opportunity to reflect. The combination of an inside-the-classroom view combined with the perspective of a profession outside the classroom offers the chance to see what may be invisible from the trenches of daily teaching.

The visible classroom is the one in which there is a narrow focus on the task of imparting information to students. Pressures external to the classroom, like accountability testing and in some cases technological innovations like online learning, encourage us to keep a narrow focus on student academic achievement. Relational, emotional, and behavioral concerns seem like distractions. However, whether or not we see and recognize them, the hidden emotional, relational, and neurological factors at work every moment among the faculty and students wield a powerful influence on what will actually be learned.

The term *invisible classroom* refers to the web of neurological and human connections that create the context for teaching, learning, and living. As the boys taught me during my first few weeks of teaching and an immense amount of research since then supports, a classroom is far more than a room of separate individuals; it is a web of interconnected relationships, with forces operating with their own rules that change the neurology of all involved. We may not realize it, but we educators are always interacting with these forces, in ways that can support or unintentionally hinder learning, making our teaching less effective, more stressful, and less fulfilling.

On my first day in that high school class, I unintentionally ran afoul of the invisible classroom. As the semester unfolded, the boys became more connected with each other as they united against their common enemy (me). It crippled my teaching and the students' learning. Now thanks to research in the areas of relational neuroscience, positive psychology, and mindfulness, educators can do much more to avoid mistakes such as mine and more—we can harness these invisible forces to support teaching and learning.

Standing before a classroom every day is undeniably a courageous act. You, as teacher, stand where the hopes and dreams of adults—be they parents, politicians, or administrators—meet the reality of a group of young people with diverse abilities, interests, and personal histories. The task at hand is nothing less than shaping young brains for success in the world, and ultimately their ability to develop and contribute their talents will determine the sustainability of our world. You deserve whatever support is possible to make this endeavor both effective and fulfilling. This book is a humble attempt to provide that by making the invisible aspects of a school and classroom visible so you can address them in ways that enhance learning and improve teaching, while feeling fulfilled by your critically important profession.

Foreword

A QUARTER CENTURY AGO, INTERPERSONAL NEUROBIOLOGY WAS BORN in the hearts and minds of clinicians and researchers from around the world. Psychotherapists, neuroscientists, developmental psychologists, and physiologists began to integrate empirical and observational data in an attempt to understand human experience through the consilience of multiple scientific and experiential lenses. We benefited from the work of the women and men who came before us who explored the worlds of attachment, social psychology, and learning theory. Along the way, those in the field of social and affective neuroscience, evolutionary theory, and epigenetics joined us. The results of our studies were primarily directed toward psychotherapy, parenting, and child development.

Over the years, the application of interpersonal neurobiology to classroom teaching was both obvious and unexplored. After all, both the consulting room and the classroom are learning environments that depend on the same biological, emotional, and social forces to stimulate learning. For some reason that is still unclear, few educators participated in the early development of interpersonal neurobiology despite its obvious relevance to learning of all kinds.

When I turned my attention to education a few years ago, it became clear to me that interpersonal neurobiology held the key to improving learning outcomes and contributing to a better world. In my book, *The Social Neuroscience of Education: Optimizing Attachment and Learning in the Classroom*, I began to lay the groundwork for a new science of education based on how we have evolved to learn—in the context of secure attachment within caring relationships.

In tribal societies, education was part of everyday life. Within tribes, our brains evolved to learn in the context of naturally occurring apprenticeships with closely related others. This accounts for the natural synthesis of attachment, emotional regulation, survival, and learning. When education was converted into a process of industrial mass production, the attachment, emotions, and apprenticeships were replaced with large classrooms, competition, and test scores. When the humanity is removed from education, so is much of its effectiveness--especially for those who suffer with insecure attachment, trauma, and other challenges.

In American education, what is most important about education has become invisible. If it is true that to a hammer, everything looks like a nail, then to modern educators, everything must be directly relevant to test scores. The human process of education has become a shadow, largely ignored by our educational systems. Gangs, on the other hand, who offer attachment, safety, and soul are doing better than ever. People in gangs are joined together by common interests and brought even closer by common enemies. These are the physical and social factors to which our brains have adapted for countless generations.

In *The Invisible Classroom*, Kirke Olson has built upon these foundations and the work of the central figures in interpersonal neurobiology—Siegel, Porges, Schore, and others. Combining his academic study with his decades of experience in the field of education, his goal is to forge new pathways in the application of social and emotional neuroscience into the classroom. Although he does an excellent job in summarizing and synthesizing much of the relevant research, the real gift of this book for me are the stories Kirke shares from his years of guiding students in their journey to becoming successful learners.

The invisible classroom is the matrix of social and emotional connections that serve as the tribal glue of classrooms, schools, neighborhoods, and communities. Ignored by policy makers, administrators, and most teachers, these factors make up the core of a student's experience. And, as we have come to understand, modulate the brain's ability to learn. Kirke's mission, like Freud's, is to make the invisible, visible. He shines a light on the processes of attachment, safety, emotional regulation, and implicit memory to help us understand the inner emotional world of students. He also helps us remember to look beyond and beneath disruptive behavior to its symbolic meaning for the student: anxiety and fear, trouble at home, or worse.

The expansion of the academic curriculum to include an understanding of the brain and mind is central to Kirke's recommendations for the classroom. He reviews the benefits of mindfulness exercise that include positive effects on emotional and behavioral self-regulation, increased attention and decreased anxiety. Interwoven with the science and experience Kirke presents in each chapter, he also provides classroom teachers with suggestions for direct application during their day-to-day interactions with students.

Kirke uses social neuroscience not only to help teachers understand what is happening within the inner worlds of their students, he also provides us with ways that we can teach our students about their brains. In the pages ahead you will learn that as students come to grasp how their brains work, they are better able to use their minds to cope with stress, regulate difficult emotions, and direct their attention in the service of learning.

Focusing on assessing student growth on social emotional skills, such as theory of mind, empathy, and mindfulness, may be needed to balance the current system's focus on test scores. If assessment is put into place, and students and teachers are evaluated on these essential human abilities, it may help to counterbalance the potentially destructive hyper-focus on academic test performance to the neglect of human growth.

One of the negative side effects of an emphasis on test performance is a focus on weaknesses—mistakes, missing knowledge, and the number of points away from perfection. Counterbalancing these innate problems that can be destructive to motivation, enthusiasm and exploration. Assessment of social emotional skills can have an emphasis on what the child has achieved and support the kind of ego development that is especially important with navigating the challenges of using one's intelligence to accomplish a happy and successful life. Kirke's focus on the application of positive psychology and mindfulness and their application to the classroom serve as steps in this direction.

The reality that we need to pay attention to more than the content of curriculum and test performance is an idea whose time has come. We have to learn to see and pay attention to the dark matter of the classroom. That which we are not accustomed to looking for, but which holds the educational universe together. Teachers and administrators can use this book as a blueprint to begin to make the invisible visible and to turn our human investment in our students into improved learning outcomes. Enjoy the adventure that Kirke has created for us into the inner experience of education and learn ways to transform classrooms and schools.

Louis Cozolino
Series Editor
The Norton Series on the Social Neuroscience of Education

Introduction

SCHOOL IS THE ONLY INSTITUTION THAT TOUCHES EVERY PERSON IN this country. Teachers have been influencing the development of students since the creation of formal classrooms, possibly by the Romans in the middle of the fourth century BCE. Over this long span of time, we have discovered a host of methods for passing on information to our students: lectures, Socratic questioning, classroom discussions, project-based learning, readings, homework, slide film and video presentations, and more—tools the visible classroom uses to transmit knowledge. In addition to these methods, many educators have a vague sense that there are other forces at work in their classrooms. Some of us instinctively use them to support and improve our teaching; others, myself included, have ignored these forces and suffered the consequences.

There are a host of dynamics operating below the surface of the visible classroom that have a strong influence on who learns what, when, and how. These factors are invisible but will cause trouble if they are ignored. For example, a teacher might have to deal with disruptive behavior caused by a student's pain and anger at the lack of connection to others, anxiety created when students feel unsafe, learning challenges worsened when students feel defeated by high expectations without emotional support, and inattention increased by anxiety from events outside of school—to name only a few. These troubles are signs that the invisible classroom needs our attention. They can be alleviated once we can see what has been hiding in plain sight all these years.

The *invisible classroom* refers to the microscopic neural connections inside all of us and the hidden human connections among us. These webs of neurological and interpersonal connections create the context for teaching, learning, and living.

We can uncover and positively influence the invisible classroom for the benefit of our students, ourselves, and ultimately the whole of society by using the evolving principles from interpersonal neurobiology, the study of the way relationships, mind, and brain interact to create our mental lives (Siegel 2012b); positive psychology, the study of what is best with people and what goes right in life (Seligman 2002; Peterson 2006); and mindfulness, a simultaneously ancient and modern practice (Kabat-Zinn 2005). We don't have to wait for trouble to arise before acting. We can prevent trouble while we intentionally develop a better working and learning environment. It is not too much to say that learning to see the invisible classroom will help you begin to alter the formerly invisible forces and create an environment that supports learning. All of this is as true in kindergarten as it is in postgraduate study.

My goal in this book is to provide practical understandable principles drawn from research and experience, illustrate them with true composite stories drawn from the real imperfect world of education, and offer realistic guidance for applying them in the midst of the ongoing pressures of teaching. Hopefully the students in these stories will seem familiar, although their lives may appear more difficult than that of an average student. My role as psychologist allowed access into people's lives unavailable to other professions. The hidden details and sometimes unsettling aspects of students' lives illuminated in the stories may not have been available to educators, but the principles illustrated by them are universal. We are all part of the web of interpersonal connections, and we are all blessed with the connections inside our brains.

Neuroscience: Begin with the brain. Teaching changes the structure of the brain in far more complex ways than any brain surgeon's scalpel. We educators don't need to study surgery to be able to teach, but a working knowledge of some basic brain processes helps us understand the puzzling aspects of some of our students and generally improves our teaching. There will be brief pertinent, understandable descriptions of brain functions as needed throughout this book.

Relationships: What's love got to do with it? Extensive research over many decades shows that from the first moment of birth, human brains are wired to learn best within the context of loving relationships. This does not change because children enter school, so cultivating a positive relational culture in your classroom and school supports learning and creates a better working atmosphere for you. I delve into the research on human connection so you can see how to apply it in the classroom with students and among staff. What's love got to do with it? Well, as you will see, everything!

Lead with strengths. Education focuses on discovering what students do not know, and then teaching it. This perspective has an unintended downside in that it can lead educators to minimize or ignore student strengths and passions. For educators and students alike, discovering and improving strengths leads to excellence and well-being. There is a growing body of research that supports purposely helping students find their strengths and shows us how to do it.

Anchor with mindfulness. Mindfulness is simultaneously ancient and contemporary. Its positive effect on the brain makes it a useful daily approach to the classroom. Brain research clearly shows the benefit of mindfulness, and I discuss how to use it for yourself as well as in the classroom.

The research results we will explore are not obscure, isolated, ivory tower facts; they translate into down-to-earth, practical methods that have a consistent positive effect on students, staff, and parents alike, whether the students have daunting challenges and lives filled with tragedy or lives of normality or even privilege. The examples included herein are often drawn from experiences with difficult-to-teach students in various settings. Some of the settings are plagued with rural poverty, some are from more affluent suburbs. The state of mind when teaching these students, however, is applicable in any setting, with any student from kindergarten through university.

In Chapter 1, we begin by revealing the invisible aspects of our classrooms. Beginning with the brain, we discuss how learning about love, leading with strengths, and anchoring with mindfulness provide practical direction for creating a superb learning experience in your classroom.

Chapter 2 takes us into a consideration of the human nervous system's workings to investigate why it is critically important for our students and ourselves not only to be safe but to *feel* safe in our schools and classrooms. Neuroscience tells us that without this foundation, teaching and learning are dramatically impaired.

Building on the foundation of safety, in Chapter 3 we explore why human connection is necessary for learning. Far beyond a nicety of the classroom experience, humans are wired from birth to connect with each other, and this does not change when students enter school. When teachers and administrators attend to the human connections in classrooms and schools, people can flourish; when they don't, people can languish.

Chapter 4 explores the brain's top-down and bottom-up attentional circuits and how to use them to capture student attention and guide their minds toward learning.

Memory, the foundation of all learning, is the subject of Chapter 5. The chapter begins with an exploration of the differences between explicit and implicit memory and how they can either support or undermine each other. The discussion moves on to how memories are encoded, stored, and retrieved; how there is a complex relationship between stress and memory; and how adjusting our understanding of why some of our students struggle to remember information can help ameliorate their challenges.

Growing out of this brain science–based theoretical and practical foundation, Chapter 6 draws from positive psychology, the study of what is best in people, to explore the positive effect of balancing, helping students improve their weaknesses with nurturing their strengths. We will learn how and why a strength-nurturing approach can increase our students' learning as well as bring respect and connection to administration, faculty, parents, and students alike.

Practical applications of mindfulness, the nonjudgmental attention to the present moment, as a means for helping our students (and ourselves) find our center and open to new learning and warm relating is the subject of Chapter 7.

We put it all together in Chapter 8 as we circle back to the beginning, weaving the principles and practices of relationships, neuroscience, and mindfulness into the possibility of improving school and classroom culture.

In my experience, this approach works best if it begins at the top with administration, staff, and faculty practicing together the changes they would like to see with students. Often, these experiences alone begin to influence how we perceive our students, and the classroom culture begins to shift even before we directly apply these new methods. My wish for educators is to find the deep satisfaction in the critically important hard work we do, side by side with our students settling into meaningful rewarding lives as they make positive contributions to our world in the years ahead.

Chapter 1

Uncovering the Invisible Roots of Learning

MY PICKUP TRUCK COMES TO A CLATTERING STOP IN THE SCHOOL parking lot. Through the windshield, I see small groups of students lingering outside in the morning sunlight. Some are chatting with each other; some talk and laugh with the principal; a young couple stands separately, showing the first signs of blossoming love; others are shooting baskets as they tease each other. I notice two boys running awkwardly toward my truck. Tucker is informally dressed in dirty overalls and work boots, Jimmy wears a sweater over a white shirt. Both are smiling.

Tucker (completing his morning ritual) laughs, yanks open the passenger door, and hops in: "I love your truck. Will you give it to me?" As I give him my daily answer and shoo him out, Jimmy, standing in his characteristic awkward stiffness, excitedly informs me that the insect order *Lepidoptera* (moths and butterflies) was named by Carl Linnaeus in 1735. He asks, "Do you know about all the species in that order?" I refer him to his biology teacher, who is also arriving, and he dashes to her car.

As I walk toward the school building filled with students placed here by the special education system because they could not be taught in other settings, I take a moment to reflect. How can there be so much joy and warm human connection among students who some believe are unable to connect and others consider simply unteachable? Why, with no formally assigned bus duty, are so many among the staff standing outside warmly greeting students every morning (even during the cold, dark days of winter)? How is it possible for Tucker, who last year was chronically angry and even injured a teacher in his previous school, to be laughing and joking with the school principal? For that matter, what happened to change the boy excited about *Lepidoptera* from someone so afraid he could not enter a school building without vomiting into a student who feels safe enough to eagerly greet staff in the morning?

The daily challenge of teaching students after many talented educators had tried their best and failed pushed me to look for potentially helpful research results from fields outside of education. I examined areas such as interpersonal neurobiology, positive psychology, social neuroscience, mindfulness, organizational development—anywhere an answer to the puzzle of students' challenges might be hiding. Because the research included infants, children, adolescents, adults, animals, organizations, and businesses, it has useful applications for all of us: students, teachers, administrators, parents, classrooms, and schools alike. It seemed the most helpful results could be roughly grouped into discoveries about neuroscience, relationships, strengths, and mindfulness.

Beginning with the Brain

What can recent brain research teach us about the roots of learning? In 2001, I began to be fascinated by neuroscience, particularly interpersonal neurobiology (Siegel 1999, 2012a)—the study of how our relationships continually shape our brains. A good starting place for schools and teachers is to dedicate ourselves to learning about the core circuitry of the brain—not just the cognitive circuits but the relational and emotional circuits as well. This knowledge can be applied to understanding how teaching and learning can be enhanced by the quality of our relationships with each other, and thus we can create a school culture that supports excellence. In the process, we can begin to understand that students have patterns neurobiologically ingrained from their early experiences that can, for example, make it difficult for them to imagine school as a positive experience. In the case of difficult-to-teach students, I began to suspect that this negative perception of school might be adding an impossible load to the burden of their learning challenges.

For an example of how interpersonal neurobiology can inform our teaching, let's take a close look at Tucker, a student who came to my school after his previous teacher lost her temper and yelled at the school counselor and special education director, "I don't care how you do it, get this kid and his mother out of my class!" Tucker was physically large for a seventh grader; he saw no use for school; he hated reading, history, and math; sometimes he was interested in science; he loved to hunt and fish and often spoke of his father's gun collection and their hunting excursions. At home, he blamed the teacher for his poor grades. His parents had attended the same school when they were children, had hated it, and always believed Tucker's negative stories about his teacher and school.

How can neuroscience help us get Tucker (or any student for that matter) into a state of mind open to learning? In the late 1940s, psychologist Donald Hebb (1949) discovered a simple but profound neurological process. Anything we experience causes a network of neurons to fire. Imagine these complex networks of neurons that can span the brain and the body as resembling a fishing net. Like a fishing net, each fiber is connected to every other fiber to create a whole net.

The more frequently the same net of neurons fire together, the more likely they will fire in the same pattern again. This is summarized in the phrase, "What fires together, wires together." This simple but profound process applies whether we are learning course content, discovering how relationships work, living our daily lives, or developing a sense of how school will be.

To understand how this neuroscience process can help you with your students, take a moment to imagine yourself in Tucker's life. As a young child, he heard his parents complain about school and tell stories of their negative school experiences. Every time he heard the complaints, a network of neurons fired in his brain, and the more often they fired, the more likely they would fire again in the same pattern. Without realizing it, his parents were creating an ingrained neural network in their son that led Tucker to expect that these same negative experiences would happen to him when he went to school.

Let's give this complex web of neurons the crude yet accurate name of the "school sucks" neural network. Over and over, his parents' stories unintentionally made the "school sucks" network fire and strengthen. Imagine Tucker on his very first day of school, with his wired-in "school sucks" network at the ready as he enters the building. Without his conscious knowledge, Tucker's brain is expecting negative experiences at school. In the multitude of small experiences that make up a day at school, he will certainly find some negative ones. If nothing else, he will interpret his normal first-day anxiety as supporting his "school sucks" neural net. Imagine him arriving home after his first day, saying something like: "That place is creepy. Those big kids are scary." His parents, triggered by their son's fear, share negative stories from their past at school. Tucker's "school sucks" network triggers his parents' "school sucks" network and they strengthen each other. If nothing changes, the next day will be a repeat, further strengthening the negative network until the thought of school produces a feeling of dread and resistance through his whole body—hardly an ideal state for learning.

The thought of breaking through this kind of negative neurological cycle might seem hopelessly daunting. But this is far from the case because of another neuroscience discovery—*neuroplasticity* (Siegel 1999, 2012a; Schwartz and Begley 2003; Doidge 2007). Simply put, neuroplasticity is the brain's ability to change its pattern of neural connections throughout the life span. Neuroplasticity is a discovery from neuroscience that offers educators much hope, because unlike the outdated belief that the brain is static after about age seven, neuroplasticity shows that it is possible for anyone to change their neural networks with some effort and with support.

How Neuroplasticity Can Work for You

No matter how much you may want to, you cannot go inside Tucker's head with a pair of tweezers and disconnect his "school sucks" neural network. However, you can help students change old neural networks by creating more positive relational experiences for them at school. The change can begin, for example, simply by greeting each student individually, authentically, and warmly, allowing him or her to feel felt and seen by you. Taking an interest in his or her strengths and passions, no matter how different they are from yours, will let each student know you value him or her. Incorporating some aspects of mindfulness practice (described later) will help calm the fear that may be activating some of these neural networks. As these positive relational experiences are repeated, the old neural network will begin to be transformed into one carrying the new experience. This process is called *memory reconsolidation* (Ecker et al. 2012). Part of Tucker's "school sucks" pattern is experienced in his body: his muscles get tense, his gut tightens, his heart speeds up at the very thought of going to school. In this state, he is primed to notice anything that is even a remote trigger of the "school sucks" network. He also has the anticipation that teachers will greet him warily, since that's what his parents described and what happened at his old school. All of this creates a state of mind that makes it nearly impossible for him to learn.

If instead, when he arrives, he is greeted personally with authentic warmth, that simple gesture will help his body begin to relax a bit. In that moment, the old network is being modified very slightly because of the difference between what his "school sucks" neural network led him to expect and what actually happened. This is called a *disconfirming experience*, an essential building block of the memory reconsolidation process. When we anticipate one thing but experience another, our future expectation can be modified slightly—sometimes a lot. By repeating these experiences daily, gradually large shifts in perception are possible. The beginning of creating a foundation for academic learning is establishing a relational environment that collaborates with the brain's natural learning processes. When kids feel safe and connected, the doors to learning open wide.

It begins for us educators by accepting students for who they are in any given moment—although we may not actually like who they are in that moment. This can be a difficult task, but anything other than personal acceptance increases fear and reinforces negative neural networks. For example, when Tucker first began attending my school, we did not try to convince him school is really important for his future or that he might as well make the best of it. We intentionally ignored these typical talking points. Instead, we encouraged him to join the outdoor education program and even invited his parents to join him. Although they did not, the invitation was likely a small disconfirming experience for them, because it was an authentic invitation by school staff—people whom they had long considered negatively—to engage in an activity they have enjoyed. The staff members were told about the family's outdoor interests, so they looked for opportunities to swap hunting and fishing stories with Tucker's parents at various school events. These experiences and others began to add a "this school is different" or a "this school is okay" stream of information to the old neural network for the family.

A good deal of effort was made to repeat these experiences so the old neural network could continue to transform. There were ongoing one-to-one discussions about his interests, encouragement in classes, and offers for Tucker to have a voice in decisions about his classes. It is possible that these efforts at school might have triggered positive school-related discussions at home, further transforming the neural network holding his perception of school.

It is important to know that the "school sucks" neural network does not just vanish immediately; it takes time for many disconfirming experiences to weaken the network, just as it took many experiences to develop and strengthen it. During this process, the network may be easily activated by perceived negative school-related events, and when this inevitably happens, it is a temporary setback. But the ongoing repetition of positive experiences—intentionally created—continue to make the "this school is okay" neural network repeatedly fire and become stronger. In Tucker's case, as this new neural network became established, he began to feel safe and accepted, making room for him to anticipate positive events at school. From the foundation of the "this school is okay" network, we could help him build other positive neural networks that supported his learning. For example, we might build a bridge to new learning in biology with a project linking his knowledge about game animals to related topics; in his history class, we could link his hunting and fishing passion to the lifestyles of the early American colonists.

Tucker is an example of how understanding and using neuroscience's basic processes and neuroplasticity gives us tools to create the kinds of experiences that have the potential to change habits of thought, feeling, and behavior that create obstacles to student learning. The details of students' experiences and their preexisting neural networks as they enter our classes vary widely. Neuroscience invites us to think about any attitude or belief that creates an obstacle to learning as a neural network that can be altered by creating positive relational experiences and disconfirming experiences for students.

Applying these and other research results is both demanding and exhilarating. In the opening chapters of this book, we consider discoveries about the autonomic nervous system and the importance of safety for learning (Chapter 2); explore the ways that warm connections prepare the brain to take in new information (Chapter 3); and learn about the neural processes of paying attention and remembering (Chapters 4 and 5)—all with many practical application ideas. At school, these have key places in the invisible classroom. How we see ourselves and our students prepares us to offer experiences that collaborate with the brain's natural learning process.

Leading with Strengths and Passions

Tucker's story alludes to a second avenue for creating a classroom culture in which learning becomes optimal: focusing on strengths and passions in addition to the remediation of weaknesses. In 2002, Marty Seligman published his seminal book, *Authentic Happiness*.

Seligman is credited with being the founder of positive psychology. He and several other psychologists realized that for over a century, the field of psychology had been answering the question: "What is wrong with people, and how can we help them improve their weaknesses and overcome their pathology?" He and his colleagues began to ask the question: "What is right with people, and how can we help them expand their strengths?" At my school, the staff took these concepts to heart, and we dedicated ourselves to uncovering and supporting expansion of our strengths and passions. We completed a number of assessments, experimented with several exercises developed by positive psychology researchers, and applied what we learned about each other during the daily life of the school. We formed faculty committee assignments based on personal strengths and passions rather than seniority or other reasons. For example, one teacher was passionate about role-playing games, so she began an after-school gaming club that has been very popular. We were so heartened by the results for both the quality of our work and the improvements in mood and relationships among the staff that we began to turn our attention to offering this strengths emphasis to the students. Here is what happened for one of them.

Gary was good-natured at home but would explode in anger at school and was barred from returning to his previous campus after injuring a teacher during a classroom fight. He arrived with his parents at an admission interview for our school surly, sullen, and angry. He appeared more prepared for a fight than an interview. His parents were shocked, saying he was fine on the drive to the school, but once inside the building he had a total change. We asked the questions we always ask first: "What are you good at? What do you love to do, in school or out of school?" He seemed surprised by the questions. We were stunned by his answer. He described waking up that morning before dawn and driving his snowmobile to the top of a local hill to watch the sunrise, returning before his parents woke up. He poetically described the reds, blues, and grays of the winter sunrise. "It is where I find peace. Other times, I just walk into the woods and sit quietly, but the colors of the sunrise from that hill are amazing."

Those first questions uncovered his passion for the natural world. A *passion* is defined here as a very strong emotional attraction to something. An *interest* suggests an attraction to or curiosity about something but with less intensity than a passion. *Strength* is defined by British educator Jennifer Fox Eades: "Strengths are capacities to think, feel, and behave in certain ways. They represent what is best about us and when we use our strengths we are energized, we sparkle and soar, we achieve the highest goals we are capable of achieving" (Eades 2008, p. 34). People pursue a strength or passion for its own sake, often during their leisure time. Strengths and passions can be interwoven (but not always). For example, many adults I know are passionate about a sport such as basketball, tennis, or golf, but demonstrate that the sport is not a strength by not playing very well.

Suspecting Gary had strengths paralleling his passion for the natural world, we asked more questions, uncovering his skill at tracking animals, identifying plants and birds, and even cooking on an open fire. As the staff in the interview began sharing their own experiences hiking, biking, hunting, and skiing, they began to develop a strength- and passion-based connection with him. The glaring fact of his violence and his upcoming court case could have easily blotted out the beauty and color of his sunrise and the uncovering of his strengths, but the strength-focused questions allowed his light to shine. I will talk more about Gary later, but the point here is that focusing on strengths and passions can begin a process that has the potential to change even violent students' lives for the better. In the process of noticing strengths in ourselves and our students, our state of mind begins to shift from a primary focus on the often arduous process of unearthing and improving student weaknesses to the explosion of energy that arises from discovering and expanding strengths and passions that are already present. (Chapter 6 explores the practical implementation of strengths-focused education in depth and offers many tools for you to use in your school.)

Anchoring with Mindful Moments

Side by side with the perspectives and processes offered by neuroscience and positive psychology, mindfulness offers tools to help students find a calm center from which to learn. *Mindfulness* is simultaneously ancient and contemporary; stated simply, it is the nonjudgmental attending to the present moment (Kabat-Zinn 2005; Siegel 2007). After hectic days in the classroom, the staff at my school began to experiment with bringing our attention into the present moment by sitting quietly and focusing on our breath for a few minutes at the beginning of faculty meetings. We then found that meetings flowed much more easily. The gatherings were more productive, and we felt better while doing it. Together, the staff decided to take the next step and introduce the students to a minute or two of mindfulness throughout the day. Over about a school year a structure evolved, where by staff and students gather each morning for a meeting that begins with a few mindful moments, and many classes begin the same way. As the mindful minutes became part of the school culture when a class become too disruptive the teacher simply had to say, "let's take a mindful moment" and the students quieted rapidly. This is the structure that works in the culture we developed in our setting, and each school and classroom can develop its own structure to add mindfulness.

Mindfulness practices build neural strength in the areas of the brain that allow us to focus and take in new information with less interference from wandering streams of thought, emotion, and external distraction. In addition, these practices build the neural capacity for caring relationships (Siegel 2007). Chapter 7 offers an in-depth exploration of mindfulness and resources for applying it in educational settings.

Building on a foundation of consciously offering safe, secure relationships for our students, we have found that the combination of applied interpersonal neurobiology and a daily emphasis on strengths, coupled with the anchor of mindful moments throughout the day, gives us a powerful trio of perspectives and interventions that can improve any classroom from kindergarten to university. At the end of each chapter, you will find Tools for School, offering practical applications for what we discuss. As you translate these suggestions into what works for you individually in your setting, I hope you will find ease and satisfaction in your valuable work.

Chapter 2

The Neurobiology of School Safety: Being Safe and Feeling Safe

MOST EDUCATORS BEGAN THINKING SERIOUSLY ABOUT CLASSROOM and school safety after the tragic school shootings at Columbine High School in 1999, not because it was the first but because that particular horror changed the world of education. In one tragic day, school went from a place of presumed safety to a place of potential danger and real anxiety. The change was abrupt, stunning, and sad for educators everywhere. In the haste to protect students, we increased physical safety, but also increased staff and student anxiety. Metal detectors, buzzing door locks, and safety officers in the schools became part of daily life, but these mechanisms of protection were also daily reminders that we needed to be protected.

Who and what did we need protection from? Outsiders attacking the school like what happened in Newtown, Connecticut, in 2012? Insiders (students) like at Columbine? Terrorists from other countries like September 11, 2001? When what was previously unthinkable became possible, potential danger lurked everywhere, making fear and anxiety part of daily life. Anxiety distorted educators' decision making, because any student could be a potential threat. School administrators asked psychologists from outside their schools to complete safety assessments on students who had done something that was common prior to Columbine, but post-Columbine these same actions had suddenly turned sinister. For example, I was asked to conduct safety assessments on a nine-year-old boy who forgot to take his jackknife out of his pocket at home and showed it to a friend at school; an angry ten-year-old girl who wrote B-M-O-B (misspelling *bomb*) on the girl's room wall; a seven-year-old boy who brought a rusty unworkable pistol he found in the cellar of his grandfather's barn into school for show and tell. All were suspended from school until the safety assessment was complete.

As these examples of the misinterpretation of student behavior demonstrate, when the neurobiology of fear is active, our thinking narrows, turns rigid, and becomes focused on the perceived threat. When the fear circuits of the brain are active, clear-headed decision making and new learning is difficult or even impossible. What is needed in addition to physical safety is emotional safety, the internal felt sense of being safe. The mechanisms of protection like locks, metal detectors, and security cameras help us and our students be as safe as possible. They are necessary, but these mechanisms are not sufficient to create emotional safety.

As we move toward the basics of a neurobiology of safety, let's consider the important distinction between *being* safe and *feeling* safe. The harsh reality is that we can never be sure our children are 100 percent physically safe; in spite of this we can and should create a school culture where students and staff feel safe so learning can occur.

The school tragedies show us that, like Maslow's (1954) hierarchy of needs, physical safety is only the beginning. As you may know, Maslow's hierarchy is usually shown as a pyramid beginning with the most basic needs at the bottom and progressing to more complex needs at the top. Physiological needs such as breathing, food, and water are at the bottom. Safety is the level above the basics that support life. Love and belonging needs are at the level just above safety, and they are followed by self-esteem needs (e.g., respect of self and achievement). The hierarchy culminates at the top with self-actualization (e.g., the process of individual growth and reaching one's full potential). Although the hierarchy is no longer considered a rigid step-by-step progression (it is generally agreed that all the needs at one level need not be met before moving on to the next), Maslow's hierarchy shows us that physical safety is a necessary part of the foundation of a school and the sense of belonging and connection, providing the additional layers that help all members of a community settle into a comfortable openness with one another.

The most basic layers of being safe are the physical measures (locks and alarms) installed for safety. Feeling safe depends on our interpersonal environment, possibly more so than our physical environment. The most basic measures for feeling safe in our interpersonal environment are the policies designed to prevent discrimination, sexual harassment, and bullying. These protections are only a notch above metal detectors; they are designed to prevent negative things from happening and instruct us what to do if they do happen. Only by intentionally and continually developing trust, tolerance, and the acceptance of differences will we arrive at a school culture that feels safe and is fully supportive of learning. Emotional safety depends on mutual trust and trust in an organization begins at the top: between administrators and teachers, and it then is communicated by teachers to their students.

The connection and attunement between teacher and student is critical for students to feel safe in the classroom. Louis Cozolino explains it well: "From the neurobiological perspective the position of the teacher is very similar to that of a parent in building a child's brain. Both can enhance a child's emotional regulation by providing a safe haven that supports the learning process. . . . Among the many possible implications of this finding for the classroom is the fact that teacher-student attunement is not a 'nice addition' to the learning experience, but a core requirement" (2013, p. 18).

Wiring for Safety

Thanks to the work of Stephen Porges (2011), we have a more complete understanding than ever before of how our autonomic nervous system (ANS) is central to our experience of feeling safe. The ANS is a control system that operates below the level of consciousness. It originates in the brain stem and has connections throughout the hollow organs (e.g., heart, lungs, stomach). It is divided into three subsystems: the sympathetic nervous system (a rapid-response mobilizing system) and two branches of the parasympathetic system, both of which are slower to respond and quiet our systems. As part of the innate functionality that ensures our survival, these systems are rapidly and profoundly responsive to our environment, including the internal state of those around us. The ANS, along with other neural circuits, is constantly scanning, in about quarter-second increments, to determine whether we are safe. Humans particularly find safety with fellow human beings, and so we attend to them to determine if they are with us or separated from us. When we have the sense that the people around us are accepting, curious, predisposed to believing we are doing the best we can (and therefore largely not judging us), and willing to be helpful, we have a felt sense of being safe. When we sense criticism or judgment, anxiety, anger, or disinterest, we feel unsafe. For example, think how your body might respond when you were openly and warmly welcomed as you entered a good friend's home. Now contrast that response to how your body might feel as you enter a room full of stern-faced well-dressed school board members.

Porges coined the word *neuroception* to explain how most of this instantaneous response occurs. It is nonconscious—outside of our conscious or perceptual awareness—yet it influences our future actions. Your neuroception of welcoming friends influences you to feel safe in your time together and will make it comfortable to share your intimate thoughts and listen attentively. Your neuroception of a room filled with stern school board members will influence your response as well, probably making you feel unsafe and guarded about sharing personal thoughts and feelings, as well as cautious while listening to theirs. In both instances, neuroception is working nonconsciously without our knowledge, but sometimes there is a sudden jolt that brings us quickly from nonconscious neuroception to the conscious awareness of danger. This shift might occur when there is a sudden loud noise in an otherwise quiet room or when an an especially intimidating school board member aggressively confronts us. If we have a neuroception of safety, our system wants to open, step closer, and engage. If we have a neuroception of being unsafe, our system is designed to tighten, close down, focus on the potential threat, and take some action for protection—to fight or flee if action seems likely to help us (e.g., avoid the intimidating school board member). Sometimes there appears to be no action that will protect us, we feel helpless and our neuroception triggers us to collapse. In these responses to various degrees of fear and helplessness, a student or colleague may lash out verbally or physically, or collapse in a heap (e.g., become unable to speak in the presence of the intimidating school board member).

What factors go into these instantaneous assessments of our surroundings? One source of input are the cues we receive from the people in our immediate environment. We sense physical safety from nonverbal signals that primarily flow from the faces, voices, eyes, and bodies of those around us. Their (and our) inner states are expressed in tone of voice, the degree of relaxation in face, the dilation of pupils and expression in the eyes, rate of breathing, and the ease of movement in the body.

The second powerful source of input comes from our previous life experience. In our early life and then throughout our lives, experiences become encoded in our memories in a way that produce the expectation that what happened in the past is likely to occur in the future. For example, if one has had a lot of fearful experiences at school, one would expect the next school encounter to have a similar flavor. The memory of past experiences "colors" responses from inside, regardless of what is actually happening in the external environment. However, as considered deeply in the next chapter, we are always on the lookout for ways to connect with others—a primary biological drive—so our system is also biased in the direction of seeking connection. If those around us are able to offer a safe haven, we have a tendency to move in that direction. This gives teachers a big advantage if we become that welcoming harbor for students.

Circuits of Trust and Safety

Developing an understanding of the details of the ANS can help us build the foundation for creating a safe and supportive environment for students so they may have a neuroception similar to entering a friend's home as they enter the classroom. The ANS has three branches that work together in a hierarchical order:

1. The ventral vagal parasympathetic branch—the default mode. When we have a neuroception of safety, the circuitry

of social engagement and optimum learning is active and remains so until we don't have a felt sense of safety.

2. The sympathetic overrides the ventral vagal branch. When we have a neuroception of danger and believe we can defend ourselves, this branch activates the fight-flight-freeze response.
3. If the danger escalates to the point that we begin to feel helpless, the dorsal vagal parasympathetic branch overrides the sympathetic. The neuroception of helplessness tells us that we are in danger of dying, so we move in the direction of feigning death to avoid death—a state in which all systems slow down so much that our consciousness begins to fade and we may collapse, dissociate, or faint.

What is described here is how the sympathetic and dorsal vagal parasympathetic branches adaptively respond to danger. However, when we have a neuroception of safety, these two branches can work with the ventral vagal to support states of play and joy (sympathetic) or states of deep rest and contemplation (dorsal vagal parasympathetic).

Our whole system has a preference for being in the ventral vagal parasympathetic state because it allows us to stay in connection with one another and co-regulate each other, so it is the human default mode. In fact, according to social baseline theory (Beckes and Coan, 2011), we have such a preference for interpersonal connection that our fear system doesn't activate nearly as much when we are accompanied by someone we trust. Our ventral vagal parasympathetic system assumes the other person "has our back." When there is this kind of interpersonal trust, our nervous systems do not have to be as alert for danger and thus have a lot more energy for learning, creating, and exploring. When colleagues and administrators support us, it makes it possible for teachers to flourish. When students feel it from their teachers—for example, experiencing an authentic calm warm welcoming at the classroom door—it can be the difference between success and failure, particularly for those who come to us with shaky nervous systems.

Help Yourself: Begin with the Ventral Vagal

Because humans prefer the ventral vagal state, starting the class by being in this state ourselves offers students an open door for calming, connecting, and settling into learning, even for those who carry a lot of anxiety and upset from outside school. In the ventral vagal state, our bellies and faces will feel more relaxed and our heart rate will be moderate because of the engagement of what is called the vagal brake that slows the heart. We have a sense of calm alertness and readiness to engage, as well as a greater ability to read students' faces. We are likely to feel playful, curious, and caring toward our students (Panksepp and Biven 2012) and their neuroception of our state influences their inner states to move in the direction of matching ours.

How can we recognize, nurture, and learn to maintain this state? Creating or working in a school culture where we can be sure that fellow teachers and administrators have our backs is one important way to help maintain this state. Another individual technique for stepping into the ventral vagal state is to focus on our breathing, allowing the exhale to be a little longer than the inhale. At the end of every exhalation, there is a natural pause before we breathe in again, a kind of tidal rhythm. The practice of focusing our attention on our breath and noticing these pauses when we are away from the stresses of school can strengthen the calming pattern so that it will be more reliable when we need it. (Additional mindfulness practices like this are addressed in detail in Chapter 7.) Remember, when our students come through the door, their brains will be scanning for safety roughly every quarter of a second, and encountering a face and body that is settled and receptive can give their day a solid beginning. As our ventral vagal state helps students' ventral vagal system come online, a classroom-wide feedback loop of vagal connection and openness has a chance to develop.

The stresses of school (and the world in general) can often give teachers and students a push toward sympathetic activation, brought on by a neuroception of danger. Heart rates accelerate, eyes widen, bellies tense, tone of voice and eye gaze change, the ability to read the faces of others diminishes, thinking loses its flexibility (nuances are not recognized), and focus narrows to attend only to information related to the perceived threat (Fredrickson 2009). The social engagement system turns down, so we can no longer co-regulate each other. New learning nearly stops. All of this sympathetic activation is adaptive in threatening conditions, but problematic in the classroom.

For humans, disconnection is felt as a threat. As we become more attuned to our inner states, we can more easily sense when we have moved into this kind of sympathetic activation, call on our breathing practice, and possibly connect to a trusted colleague to find our way back to the ventral vagal state. We can intentionally develop a personal repertoire of pathways that help us notice the shift toward sympathetic arousal, identify the perceived danger, and move back toward a neuroception of safety and the interpersonal engagement of the ventral vagal. In addition to it being vital for students, spending more time in this ventral vagal state means we will be less tired and emotionally drained at the end of the school day.

Many of our students currently face daily stresses at home and anxiety from our fast-paced crisis-oriented society, in addition to the huge amount of information they must process at school.

Consider the possibility that students may have experienced earlier traumas, and then the power and importance of offering a setting that encourages a ventral vagal state grows in importance.

The Third Branch

We should also consider the third state in the hierarchy of the ANS branches: the dorsal vagal parasympathetic, which is brought online by a neuroception of danger while simultaneously feeling helpless. The body interprets this state as a signal that we may die and begins to prepare for death by drastically slowing all systems—heart rate, breathing, attention, connection to others, and even awareness of the external environment. This is a state of collapse, dissociation, and dramatic disconnection. When we feel shamed or humiliated, we are moving toward a dorsal vagal state—collapsed chest, downturned eyes, inability to respond. This state can occur when someone is the victim of bullying. For example, early in my career I worked in a school where the walls of the school office often shook as the school principal yelled at students or teachers, shaming them for some infraction. I observed students and teachers alike silently shuffle out of his office with downturned eyes, collapsed chests, and stunned expressions. With the benefit of hindsight, I can now see their response as the adaptive working of the dorsal vagal parasympathetic branch of the ANS in the face of helplessness during these tirades. Have you seen similar responses in students who have been bullied? Victims of bullying often feel helpless because they cannot fight back against the bully, nor can they enlist the help of a teacher because they believe it will anger the bully and make the episodes worse. When they feel in danger while they simultaneously feel helpless, their normal dorsal vagal response gives them that "beaten down" look common for victims of bullying.

A person may combat the feeling of collapse with defensiveness or anger, rising into sympathetic activation as a way to escape the helplessness. Although this results in a more active state, it still prohibits connection with others and continues the pattern that makes learning all but impossible. Since school is a place where many students have already experienced shame in some form, understanding this cycle and building the capacity to respond differently, even in the face of anger and defensiveness, can offer a disconfirming experience and a pathway back to ventral vagal connection for the previously humiliated or bullied students.

People are particularly vulnerable to the dorsal vagal state if they have experienced shame, humiliation, or trauma as young children, and, no doubt, some of our students and some of us educators have. With this in our background, many of us, through no fault of our own, may fall into a helpless state more easily than do others. Having warm, safe connections among school staff provides good protection and potential healing from these earlier wounds, particularly when we have built trust by feeling safe enough to share our vulnerabilities with trusted colleagues. For students, our awareness of what may be happening can help us become the refuge they need to find their way back into connection and the ventral vagal state. Remembering that each person is seeking to remain in the ventral vagal state whenever possible can give us the confidence to remain supportive when students struggle with shame and collapse by exhibiting difficult behavior.

Flexible Responding

You have probably noticed that some students struggle with finding the calm state of their nervous systems. I have often wished that simply saying "calm down and behave!" would help them regulate, but of course it usually doesn't. When they become agitated, angry, intrusive, or nonresponsive, it may be that their previous life experience has left the circuits that allow them to calm themselves poorly wired into the circuits that are experiencing the upset. When the proper wiring is in place, these key connections create a much longer and slower processing path for new and potentially frightening information. For example, if there is a loud noise in the classroom, the longer circuit with its better connections will give some students the time they need to evaluate the level of threat—did someone drop a book, or has an intruder burst in the door? The students without these connections are much more apt to react strongly and rapidly to the noise because their neurobiology doesn't give them access to a broader evaluative perspective. They move quickly into fight or flight, and their behavior follows this message from their nervous systems. It will also take them longer to calm themselves.

A longer circuit provides for both body (ANS) and emotional regulation (see Siegel 2007 for a description of this aspect of neural integration). It also gives us a quality called *response flexibility*, which is the ability to slow down and change course away from emotional reaction and toward a more empathetic relationship with others. We become less afraid and more able to see the big picture and be aware of the needs of others along with our own. The initial wiring of these circuits takes place very early in life, and when there wasn't enough support available from parents or other people caring for a child, he or she enters school with less capacity to self-regulate. Neuroplasticity offers us hope for these students.

Fortunately, we humans can "borrow" someone else's prefrontal cortex to support our own at any stage in life. Teachers in a ventral vagal state become similar to a lending library of neurobiological connectivity for those students who may have less than optimal circuitry. With ongoing support, neuroplasticity enables these students to gradually develop the missing wiring. For example, some special educators working with impulsive, behaviorally disordered students have seen hints of this underlying process when they describe acting as if they were an "external brain" for their students. This process is usually thought of as a teacher acting as an external supporter of regulation of the internal impulses of some students. It can be quite meaningful for educators to sense that we are helping our students build and repair their brains after they didn't get what they needed earlier in life.

Illustrating the Neurobiology of Safety: Brain in Your Hand

Dan Siegel (1999, 2012b) demonstrates these interpersonal neurobiology concepts with the brain made from his hand. First, we make a fist with our thumb tucked inside—this represents our brain! From the bottom up: our arm represents the embodied parts of our brain in the belly, heart, and the ANS; our wrist represents the spinal cord as it enters the skull at the brain stem (the lower palm of our hand); our enclosed thumb represents the limbic region (roughly described as the emotional/social region of the brain). The fingers that enclose your fist represent the cortex, which is the outer surface of the brain, and the fingernails of our middle and ring fingers represent the middle prefrontal cortex (a connected group of regions that processes social information, autobiographical consciousness, evaluation of meaning, and other higher order cognition). Notice that the tips of your fingers touch the lower part of your palm (the brain stem), illustrating how the middle prefrontal cortex is connected to the brain stem's fight-flight-freeze center and, under your fingers, to the thumb's limbic region.

We can use this hand model to visually describe the effect of the perception of danger on the brain. When the brain stem (palm) is activated by an internal or external alarm, these lower brain regions take over. You can show this by slowly opening your fist to demonstrate the disconnection of the middle prefrontal cortex from the brain stem. In the quest for simplicity and dramatic effect (I'm taking some poetic license yet still being neurologically accurate): As the alarm system of the brain turns on, your cortex, "the logical thinking part of your brain," becomes quiet because in danger we need to act quickly, and the lower regions of the brain take over to help you protect yourself. You can say and do things "without thinking." This is why people say and do things when they are angry and regret it later when their middle prefrontal cortex comes back on line (slowly close your fist). See Figure 2.1 for more details. This simple yet profound demonstration can have powerful effects for students and teachers.

Figure 2.1 HAND MODEL ILLUSTRATION

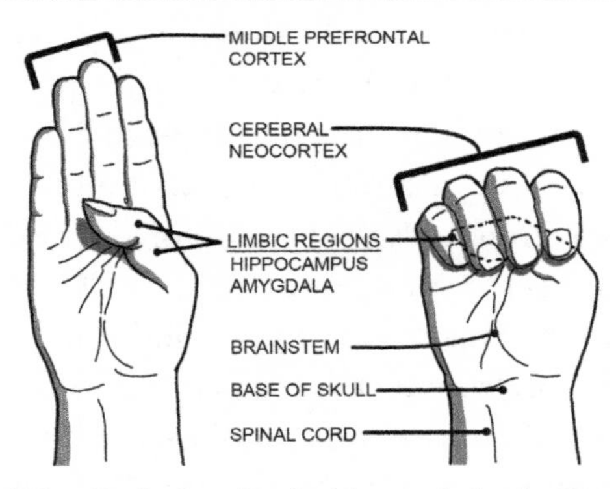

A number of years ago I described the hand model to the faculty in an after-school workshop. The next day I was walking through the school with the principal when we saw a new student in the hall. The principal said to the girl, "Tell Kirke what we talked about earlier." She rather shyly and hesitantly held up her fist in the hand model of the brain, limbic thumb tucked into her palm, cortical fingers wrapped around her thumb and said, "Today has been a difficult day—it's been a sprinkler day." Both of us were perplexed and must have shown it nonverbally because she slowly and repetitively opened and closed her fist spreading her fingers as she did. You can try it yourself now and you will see it looks remarkably like a classic lawn sprinkler. She used the hand model as a form of sign language to describe her fluctuating moods and in the process of noticing, describing, and sharing it with us, actually supported development of the very circuitry that was troubling her that day.

The hand model can be a useful tool in the day-to-day life of a school, communicating to colleagues and students in a modified sign language when the sympathetic nervous system's fight-flight response is beginning to overrule the ventral vagal parasympathetic state.

From the Top Down: Be the Change You Want

The brain's perpetual scanning to assess safety and the nonconscious working of neuroception means that educators must work diligently to keep the ventral vagal parasympathetic branch of the ANS active and return to this state when they get thrown off. When it begins to dissolve and the sympathetic system comes online, less and less learning can happen. In other words, for learning to take place, it is important to keep the fist on our hand model closed. At this point you might legitimately ask how you can possibly accomplish such a feat

Creating a school culture of felt safety and trust begins with the teachers and the administrators, not the students. If teachers and administrators do not feel safe with and trust each other, it is neurobiologically difficult or maybe impossible to create an authentic, safe, trusting culture for our students. "Do what I say, not what I do" is neurobiologically problematic when we want to develop a culture of trust and safety. In Chapter 3 we discuss mirror neurons, but for now I will just say that our nonconscious, nonverbal communication with our students powerfully influences whether their neuroception assesses us as trustworthy and feels safe in our classroom. Even more important to understand is what we communicate to our students by unconscious nonverbal means springs from our inner state. Trust and sense of safety radiates nonconsciously from us to our students, and feeling safe and trusted at school ourselves makes it more possible to create an ongoing stream of welcoming safety for our students.

Trust, Vulnerability, and Daring Greatly

A brief sojourn into Brené Brown's (2010, 2012) work may be of some help in understanding these concepts. Please note that the discussion focuses on the inner state, action, and attitudes of the people at the top, and less on their policies, announcements, and directives. People in the positions of superintendent, principal, dean, or headmaster are models for the staff as certainly as teachers and parents are models for children.

Brown has been investigating the nature of trust through research and personal experience over the past few years. She says, "Trust is a product of vulnerability that grows over time and requires work, attention, and full engagement: Trust isn't a grand gesture—it's a growing marble collection" (2012, p. 53). You are probably familiar with the marble jar as a way to encourage positive behavior in the classroom; in case you are not, this simple technique starts with an empty jar. When children behave well, the teacher adds marbles to the jar; when they misbehave, she removes marbles. The level of marbles in the jar ebbs and flows and once it is full, there is a celebration. Trust is like the marble jar—each small act adds or sometimes subtracts from the total. How do we develop trust in a school? Brown summarizes the paradox well: "We need to feel trust to be vulnerable and we need to be vulnerable in order to trust" (2012, p. 47).

We might be tempted to imagine that developing trust requires us to be in some kind of impossible-to-attain state of complete agreement among all of the members of a school staff, but this is far from the case. To get a better sense of what this means, let's consider that in the definition of trust: "reliance on the character, ability, strength, or truth of someone" (*Merriam-Webster Free Dictionary* 2013). There is no mention of agreeing with the people you trust. A robust discussion among professionals who bring into the debate diverse characters, abilities, strengths and personal truths can increase the sense of trust among them because the common focus on student learning takes competition and defensiveness out of the conversation, leading to a mutual felt sense of safety.

If vulnerability seems antithetical to leadership, I refer you to two of the most well-known writers and thinkers in the business leadership world: Robert Greenleaf and his articles and lectures on servant leadership from the 1960s and 1970s compiled into book form (Frick and Spears eds. 1998-1977) and any of Peter Drucker's many books. Using different terminology, they discuss how leadership and vulnerability meet. Whether it is Greenleaf's frustration with educational leadership in the late 1960s (Frick and Spears eds. 1998) or Drucker's (2006) advice on managing ourselves, you will see that effective leadership begins with the internal qualities of openness, listening, and receptivity in the leader, not on manipulative techniques to bring others into line with his or her agenda.

Trust from the Top Down: A Lesson from Chicago Public Schools

The importance of trust within schools has been demonstrated on a broad scale by the research following the Chicago School Reform Act of 1988. This immense restructuring and reform initiative turned the centrally managed Chicago Public School System into a decentralized system that increased local control. Bryk and Schneider (2002) directed an impressively in-depth, six-year-long research study of twelve elementary schools and their communities involved in the reform initiative. The study included several hundred lengthy interviews and observations of teachers, principals, and parents annually for three years, with follow-up surveys over the next three years. As one might expect in such a broad initiative, some schools improved greatly, some did not improve at all, and others occupied the middle ground. The authors tried to ascertain the differences between the high-performing schools and the low-performing schools. As their study unfolded, they noticed that trust among teachers, parents, and administrators seemed to be a critical factor. After analyzing their data, they arrived at a theory of relational trust that went far to explain the differences between the high- and low-performing schools. Their findings are instructive for educators on the importance of building an environment of trust and felt safety for students and ourselves.

The theory of *relational trust* begins with findings from behavioral research and philosophy suggesting that complex webs of long-term social exchanges are basic to the operation of schools (Bryk and Schneider 2002). Effective social exchanges depend on trusting relationships among the members of a school community: teachers, administrators, students, and parents. To understand relational trust as explained by the Chicago study, it is helpful to first understand a common form of trust at work in our daily lives called *contractual trust*. When you buy a product or service, there is an implied contract of trust: you pay money and receive a product or service that you trust will function as promised. Usually you can easily determine if the product functions as promised, or if a service you purchased followed an agreed-on process. There are even legal means to enforce contractual trust when a product or service is less than what was promised.

It is understandable that many people might want to use contractual trust to determine whether a teacher or school is delivering the promised product of student learning or the promised service of effective teaching. This is a mistake because contractual trust depends on reliable and accurate measurement, which is proving to be more difficult than expected for student learning, as one can see with the controversy surrounding the testing mandated by No Child Left Behind. Also, in spite of the extensive research on determining what the service of effective teaching looks like, it remains a mystery. When trying to describe good teaching, one is reminded of Supreme Court Justice Potter Stewart's famous quote (paraphrased here): "I know it when I see it."

Rather than relying on measurable variables, relational trust depends on the moment-to-moment flow of trustworthy connecting as individuals fulfill their roles with integrity, care, and in a respectful interdependent fashion. Teachers, administrators, students, and parents each play roles within a school community, and each is dependent on the others. For example, although a teacher may be seen as having more power than a student does, the teacher is dependent on students' learning and good behavior to be perceived as effective. Likewise, school administrators with seemingly more power than teachers are dependent on teachers' good performance to be judged effective school leaders. The dependencies create feelings of vulnerability among the members of a school community, complicating the situation and increasing the need for mutual trust. The researchers discovered that personal actions that lessen the feelings of vulnerability increased relational trust and contributed to the functioning of the high-performing schools. They found a dynamic interrelationship among four factors to be the root of relational trust: competence, respect, personal regard for others, and integrity.

Competence is how well an individual fulfills his or her role, be it the role of teacher, administrator, student, or parent. Each role has a visible and invisible job description and the better each person fulfills his or her role, the more the people in the other roles can trust them. Teachers work the best they can to teach, parents help with homework and make sure students arrive at school on time and rested, and administrators manage the school fairly. This allows students to step into their role as learners more effectively.

Respect is a quality of valuing other people and their contributions, with the underlying assumption that everyone has something to offer. The many long-term social interactions within a school depend on mutual respect, with each role showing respect toward members in the other roles. This includes when people disagree. It is critically important that people who inhabit each role actively listen to others: for example, administrators actively listen to teachers, teachers to parents, and everyone to the students. Because of the hierarchical nature of the school culture, respect is the necessary antidote to one role feeling overpowered by another. When trust becomes our personal default state, others can feel safe enough to risk vulnerable exchanges a necessary step to increase the level of trust among the members of a school community.

Personal regard for others allows us to feel seen as human beings as well as individuals with a specific role. Relational trust increases when members of one role see another going above and beyond their role description, because this communicates personal regard as well as respect. For example, in one elementary school, the principal covers recess duty so the teacher assistants that supervise the playground can meet, receive training, and have breakfast together. It is easy to see why the teacher assistants feel that the principal demonstrates respect and personal regard for them when she goes beyond what is expected of someone in her role to provide for their comfort and well-being.

Integrity usually means the level of consistency between what we say and what we do. In the Chicago study, the researchers found an additional aspect of integrity: "In the context of schooling, when all is said and done, actions must be understood as about advancing the best interests of children" (Bryk and Schneider 2002, p. 26). When cooperation at this level arises between administrators, staff, teachers, and parents, creative solutions to difficult problems emerge naturally within an environment of trust and safety.

Neuroception and Trust at School

Let's look at a few examples to make these ideas more concrete. You may remember that Tucker, introduced in the first chapter, had good reasons to have a highly active "school sucks" neural network. His science teacher, Ms. Atkins, has become concerned about another student in the class, Jimmy. She knows Jimmy has a history of being afraid of school, and thus has a very sensitive "school is terrifying" neural network. When he started at the school, he could not enter the building. He would pace outside, repeatedly walk toward the door, shake his head, and walk rapidly back onto the lawn. He spent the first few days unable to quiet his "school is terrifying" neural network so he could enter the building. What came next demonstrates the key to moving from a sense of danger to a feeling of safety. An administrator quietly paced with him outside the school building, often saying nothing. This enabled a process of nonverbal, nonconscious communication to occur. This happens in micro-seconds and is transmitted from body to body (Schore 2012). After several days, Jimmy was able to enter the building, sit in an office, talk with the administrator, and then gradually, one by one, join classes.

Ms. Atkins remembered all this had occurred about a year previously, and also knew that science was Jimmy's strength and his favorite class. His current passion was the family *Lepidoptera*—he spoke endlessly and sometimes annoyingly about butterflies! Recently, something about Jimmy had set off Ms. Atkins's inner alarm, a neuroception of danger that she could feel in her gut but not identify in words. She worried he was having a setback and his school phobia—his "school is terrifying" neural network—was being reactivated. As Ms. Atkins's well-integrated prefrontal cortex began to be conscious of the neuroception signals and tried to make sense of them, she began to watch Jimmy more carefully. She realized she felt a connection with him, maybe because like her, he was physically small, enjoyed school, and showed it by enthusiastically asking and answering questions during classroom discussions. However, she also noticed that lately he seemed startled when she called on him; more puzzling still, he was unable to answer questions to which she was sure he knew the answers.

One day, she noticed that as he entered the classroom, he furtively scanned the room as if he were looking for a threat. He took a seat close to her. She had an especially intense sensation of danger when Tucker entered the room and sat directly behind Jimmy. She did not see anything unusual occur between the boys, but her ANS was on high alert. As lunch period approached, Jimmy became unusually restless; his face was flushed, and he leaned forward in his seat, away from Tucker. As the classroom emptied and the students headed to the lunch, Jimmy lingered. When they were alone, Ms. Atkins took the opportunity to mention that he seemed anxious and asked if anything was wrong. There was a long, heavy silence. Tears filled Jimmy's eyes and he haltingly described the angry belittling he had been enduring from Tucker, usually at lunch. As the anxious words tumbled out, he described how Tucker made fun of his love of school, calling him a "suck up" and other things he refused to repeat. Jimmy was afraid of Tucker and could not understand why he was so angry. He described Tucker's red face, his hateful growling voice, and his fierce-looking eyes. Jimmy saw Tucker as dangerous. Ms. Atkins knew Tucker's story and did not see him as dangerous, but she could see how Jimmy's enthusiastic love of school could trigger Tucker's "school sucks" network and his anger. Finally, she could make sense of her own neuroception.

All three—Jimmy, Ms. Atkins, and Tucker—are examples of the ANS at work. Jimmy experienced Tucker as dangerous, and his neuroception turned on his sympathetic nervous system, causing him to scan the room and look for potential threats. The shift in blood flow to his voluntary muscles caused him to have a flushed face, and he sat near the teacher to protect himself. This sympathetic activation also made it hard for him to think of anything else, so he was initially startled when Ms. Atkins called on him and then could not answer her questions. For Tucker, Jimmy's enthusiasm for school was such a direct affront to his "school sucks" neural network that it enraged him. Because he and his whole family hated school, they had not been able to help him develop the neural connections between the prefrontal and subcortical regions (particularly around this issue), so he didn't have much response flexibility, and he lashed out. He had not been at the school long enough to have many disconfirming experiences to weaken that network and not enough positive experiences to build up a solid "this school is okay" network. These things take time—sometimes months, sometimes a year or longer. What Tucker did not know was that Ms. Atkins was about to design such an experience for him.

As is typical of excellent, experienced teachers working within a school culture of connection, trust, and safety, once she heard Jimmy's explanation, Ms. Atkins was able to simultaneously understand Jimmy's and Tucker's points of view. A possible resolution for the situation came to mind. She decided to invite both boys to join her for lunch in her classroom right now, while the opportunity presented itself. (She also realized if she waited until tomorrow, Jimmy could have so much anticipatory anxiety that he might not be able to come to school, and Tucker might take the opportunity to further belittle him.) She invited Jimmy to have lunch with Tucker and her in the classroom. He instantly became scared at the thought of having lunch with his tormentor, but begrudgingly stayed in the classroom because he had enough trust in Ms. Atkins. She left Jimmy digging his lunch out of his backpack and went to find Tucker. On their walk back to her classroom, she explained the situation to Tucker—who became scared and angry. He thought, "Will I be suspended and sent home again—just like my last school?"

Ms. Atkins, knowing both boys' stories and their understandable sympathetic activation around this situation, calmly brought out her lunch, offered the boys the chips she had brought for herself, and started talking. She began by saying, "Both of you have something very important in common. What do you think it is?" They looked puzzled and remained silent. "You both had really rough times in your previous schools." At that comment, the boys looked down and slowly nodded. "One of you learned to hate school; the other one learned to be afraid of school. Jimmy, you have been here longer and know you can talk with me and we will try to figure things out together. What you told me was the right thing to do, and was not ratting out Tucker. Tucker, you haven't been here long enough to know you are not in trouble like you were in your last school. You are not going to get suspended."

Tucker looked like he did not believe her, and she went on to say, "I can imagine how Jimmy's excitement about science and his love of school could drive you nuts." Tucker nodded and said, "If he doesn't stop talking about butterflies I'll . . ." Ms. Atkins gave him a stern look. "Umm, Just kidding, but . . ." Jimmy gave a shaky and barely audible response, "I've heard a million times how you shot that deer last year."

Picking up on this theme, Ms. Atkins went on, "You both have something else in common. You can't stop talking about the things you love. Maybe I'll start talking about the amazing Periodic Table of Elements again." The boys groaned in unison because Ms. Atkins's lectures about the periodic table were legendary. The neuroception of all three registered the shift in mood from fear and anger to connection with even a bit of playfulness, and they spent the remainder of the lunch period talking about an upcoming hike, something that interested both boys, if not Ms. Atkins. Her acceptance of both boys, her inner ventral vagal state, and her wise words communicated to their nervous systems "all is well," which quieted their sympathetic nervous systems—the vagal brake calming the heart, bellies relaxing to digest their lunch, and faces becoming more open. This neurobiological process enabled a positive outcome for the discussion.

Knowing Tucker's story from earlier, we can understand how this lunchtime experience was quite different than what his "school sucks" neural network would have expected. It was a disconfirming experience for him. His "school sucks" network was turned on by Ms. Atkins's invitation to lunch. He expected to be reprimanded, maybe even suspended, and to go home to enlist his parents' support against the school, just like in the past. While the old neural network was activated, he had several unexpected experiences—he was understood instead of judged, found out he had something in common with Jimmy, and had a positive lunch with his teacher and the victim of his bullying.

Recent research in neuroscience suggests that disconfirming experiences can weaken previously existing neural networks or traumatic memories (Ecker et al. 2012). Ms. Atkins's lunch weakened Tucker's "school sucks" neural network and probably strengthened his "this school is okay" neural network. Hopefully, when he goes home and describes the lunch to his parents, it will be a disconfirming experience for them as well.

For many teachers and administrators, this may not seem a firm enough intervention; perhaps they feel bullying of this sort should be punished more severely, believing that only through negative consequences or punishments will Tucker finally learn to stop bullying. This had already been tried in his previous school. It not only did not work, it strengthened his "school sucks" neural network because it was just one more aversive experience connected to school. He and his parents saw it as evidence that their negative perception of the school was correct. Understanding the interpersonal neurobiology of the events, however, can change one's perspective on how to change student misbehavior. What at first might seem too permissive or even rewarding changes into a teachable moment, a disconfirming experience, and a step toward a permanent ending of Tucker's bullying.

At this point, you might be wondering about the victim of the bullying, Jimmy. Won't he and his parents feel as if justice has not been done? Having a positive lunch with a bully and his favorite teacher is certainly a disconfirming experience for Jimmy's "school is terrifying" neural network and helps weaken it. Simultaneously, the lunch will increase the strength of Jimmy's new "school is safe" neural network. His neuroception of danger that preceded the lunch was calmed by the way Ms. Atkins created a safe lunchtime environment (e.g., offering the boys her chips and beginning the discussion by talking about the previous negative school experiences they shared). The experience taught him that his neuroception of danger, certainly warranted when Ms. Atkins mentioned having lunch with Tucker, could be transformed into an experience of safety and connection because both boys were able to "borrow" the teacher's prefrontal cortex, calming their nervous systems and gaining a broader perspective. His halting comment about Tucker's repetitive talking about his deer hunting meant he felt enough safety for him to speak up, however hesitantly, to his tormentor. This was the first time he had done that, so this was likely the beginning of a new neural network forming. In the future, Ms. Atkins will remind him of his comment, helping the neural network strengthen.

It needs to be said that if Tucker's bullying had risen to the point of physical contact, it might have been handled differently, probably with a meeting with his parents, which would offer an opportunity to change their neural networks. What would not happen is simply suspending him and sending him home, because that would strengthen the neural network we were trying to weaken and replace.

In this case, Ms. Atkins had a lot of information about both boys, so she could understand where their activation was coming from. However, it is often the case that a teacher doesn't have that degree of knowledge. It is important to remember that all of us, educators and students alike, have had different sets of experiences and memories, and those memories are part of what contributes to the felt sense of danger. Remember that memories are part of the neuroception that turns on the sympathetic nervous system. Keeping this in mind when intervening with students can make the difference between a helpful exchange and one that leads to an unnecessary, unhelpful confrontation and a bad day for all. It may help to consider that a confrontation probably means the student's and the teacher's ventral vagal system that supports connection and co-regulation is shutting down, and the sympathetic fight-flight-freeze that leads to disconnection is ramping up. Everyone's thinking has become rigid and inflexible—definitely not the time to solve problems. It is a time to quiet the sympathetic and reactivate the ventral vagal, likely our own first, followed by the student's. Developing calming rituals for these moments can begin to create a culture in the classroom so that students can anticipate that everyone will first regain their balance and then creatively solve problems. Let's look at how a principal used a creative calming ritual.

From Fear and Loathing to Thank You

A principal in a public high school of about 1,000 students used a profoundly simple ritual for helping students' sympathetic nervous systems calm and their ventral vagal parasympathetic system stay online even when they were angry and in trouble. He developed it through years of careful observation, well before neuroscience and Porges's research were widely available.

Not surprisingly, he had noticed that students arriving at the school office after misbehaving in a classroom were angry or afraid, with sympathetic fight-flight-freeze neuroception and activation in the lead. As students arrived in the office lobby, the principal greeted them pleasantly, gave them a piece of paper, and asked them to take as much time as they needed to write their interpretation of the incident that sent them there. Then he left them alone and returned to his office. From her desk, the school secretary monitored the student. When the student seemed calmer, the secretary alerted the principal, who, with studied nonchalance and often a smile, walked down the hall from his office to the lobby with his hand extended for a handshake. Even if still wary, the student's autonomic nervous system received an unexpected disconfirming experience, a nonverbal communication of a calm willingness to connect from a school principal. On their walk back to his office, the principal asked the student if he or she wanted to share a cup of tea from the Thermos he had brought from home or preferred a soda from his small office refrigerator. Students were universally surprised by the offer. It contradicted every neural network of anticipation for a principal's behavior when a student has been sent to the office.

Once he served the beverage to the student, making an overlong ritual of retrieving a glass or cup and slowly filling it, he began the discussion with an unrelated topic. Having been kindly redirected away from the threat, the student's nervous system had further space to calm down, find the ventral vagal state, and regain the ability to feel connected and think clearly. Only after the ritual was complete and the other topics discussed did the principal finally address the incident. He began by asking the student to describe his or her point of view, communicating his respect and openness to seeing the student as a unique individual. The whole exchange usually led to the principal and student together developing a plan for restitution or punishment. I observed the ritual on many occasions and often saw students thank the principal for their consequences.

In a classroom or school with a culture of safety, the opportunities for triggering students' sympathetic response are lessened, leading to far fewer messy situations. However, some triggering is inevitable. There are just too many students (and teachers) with too many invisible triggers. Learning a little about the nervous system may help us recognize when we have inadvertently triggered a student and caused an internal chain reaction, leading to the student's thinking becoming temporarily rigid and focused on a threat (which may be us, but also may have roots in earlier experiences). Knowing the student's problem-solving skills are temporarily compromised can stop us from trying to reason with him or her too quickly. Understanding the neuroscience can help us dampen our reaction to the student; after all, it is just the sympathetic nervous system doing its job, and we know it takes quite a bit longer for the parasympathetic system to quiet the nervous system than it does for the sympathetic nervous system to activate it. The principal knew that the first task is to help the students keep their ventral vagal nerve online and their sympathetic nervous system quiet. We don't have to offer tea or an invitation to "do lunch," because another method suited to your classroom and school can work just as well.

Safety in Vulnerability from Parents' Perspective

Purposely developing a culture of safety can even have a positive effect on parents' interaction with the school. An example of this occurred in a monthly meeting open to any parent. These meetings are informal and deceptively simple. I brought some pastries from the local bakery and made the coffee and tea for the morning meeting. The room gradually filled with parents, about half of them the mothers and fathers of new students and about half veterans of previous school years. There is no fixed agenda, but I have been leading these meetings for many years, so I am rarely surprised by what happens. In one particular meeting, the mother of a new student started the discussion with a pointed accusatory question: "Why does my son have so little homework?" It was a seemingly innocent question, but the familiar negative implication was communicated more by her derisive tone of voice than by her words. The underlying anxiety motivating her inquiry might look something like this:

1. For my child to be successful in life, he or she must graduate from a prestigious college.

2. To be accepted at a prestigious college, my child needs to graduate with high grades from an academically rigorous high school.
3. All academically rigorous high schools require their students to complete large amounts of homework that take a lot of time to complete.
4. Homework that can be completed in a short amount of time indicates that this school is not academically rigorous.
5. Because of this school, my child will not get accepted at a prestigious college and his or her life will be a failure (which will prove I'm the bad parent I suspect that I might be).

I began to answer her question by referring to Alfie Kohn's (2006) book *The Homework Myth*, which presents the research-based argument that homework is only helpful when it has a specific purpose, such as completing an independent project (there is actually research supporting both sides of the homework debate; see DeNisco 2013). An outspoken veteran parent who dared greatly to share her vulnerability suddenly interrupted me: "I remember thinking the same thing when my son first came here. I was freaked out. It took me about a year to realize they actually are trying to be sure he *learns*. Just doing a lot of homework does not mean he is learning anything. Especially when you do it for him. I know it's in his handwriting, but we all know who answers those questions. You know—when it's late at night; you and your kid are tired. You know what I'm saying, right?" (Many parents looked down, chuckled, and nodded.) "Relax—you can trust these people. He'll get more homework when it helps him learn and when he can do it without you."

I watched the new parent's face shift from anger to tears. Becoming vulnerable, she mumbles, "It's been so hard." A mother sitting next to her wrote her phone number on a napkin, handed it to the crying mother, and said, "Call me." Another quietly said, "We've all been there." The conversation shifted, and other parents dared to be vulnerable by telling the familiar "war stories" of the parents of students with special needs. Stories of kids being bullied by other kids, parents being bullied by other parents and often by their own family members, difficult meetings, poor treatment by professionals, physicians telling them their children will "grow out of it." As each trusted enough to share their stories of vulnerability and broken trust, the trust grew among them.

While the others talked, the outspoken parent whispered in my ear, "We've got to talk these new parents off the ledge. I remember when I was there and two teachers and some parents helped me get it—just passing it on." The parents could dare to be vulnerable with each other because they were part of a school culture that deliberately works to develop a felt sense of trust and safety. As the meeting drew to a close, it appeared that each parent had gained the trust of the others and increased their inner strength to face the long road of parenting a child with a disability. The meeting demonstrated how the cycle of vulnerability leading to increased trust described by Brown (2012) could look for parents.

For a quick review of the autonomic nervous system, let's look at the interpersonal neurobiology of these parents daring to be vulnerable.

Porges (2011) suggests that the interpersonal environment can support a felt sense of safety or of danger. When we experience safety, the branch of our nervous system that allows us to stay in connection with one another comes online. For humans, vulnerable connection is a sign of and supports the continuation of a neuroception of safety, so once online, a positive feedback loop begins that makes it more likely we can stick together as these parents did. This ventral vagal pathway lets us read the faces of others and allows our own face the mobility that expresses an invitation into supportive relationship. It influences the tone of our voice, our breathing rate, pupil dilation, and the ability to attend to the distress of others. Because the experienced parents have enough experience with a school culture that intentionally develops a shared sense of felt safety, they can maintain this felt sense of safety for themselves and their children. This felt sense makes it possible for them to help the new parents. They did this very effectively because they communicated verbally and nonverbally that this school is a safe place.

Any parent entering a new school after having spent many years when neither their children nor themselves have felt safe is much more likely to have a neuroception of danger coming to a school meeting. In the tone of voice of the parent who spoke up about her child's homework, the slight widening of her eyes, and her more rapid breathing, I (and likely the other parents) could nonconsciously sense the signs of activation of the sympathetic fight-flight-freeze system. No doubt this is a familiar and legitimate response to a meeting at a school, a common stressful event in the lives of parents of children with disabilities and perhaps of most parents. An educator who schedules a meeting with any parent could be alert to these signs of the activation of the sympathetic nervous system and purposely work to calm them.

Prior to coming to a school that purposely attends to the sense of safety of all its members, school contact for most parents has been unnecessarily negative: from the principal calling about misbehavior to stressful individual educational planning meetings filled with numerous professionals speaking educational and psychological jargon. It is no surprise that just arriving at a school for an innocuous parent meeting could activate the fight-flight-freeze response of a parent new to that school. Because of an accumulation of positive experiences over two years within a safe and trustworthy school culture, the veteran parent's ventral vagal system could remain online during the meeting. During those two years, the sense of safety and trustworthiness was reinforced by almost weekly phone contact with staff, interactions at several positive after-school events (including the school's soccer matches), and other positive conversations with staff and parents. Over time, the neuroplasticity of the veteran parent's brain had accumulated enough supportive experiences to change her neuroception from danger at school to one of safety and connection.

The new parent's neuroception was wired for distrust and conflict by her previous experiences. Her anger revealed the more familiar pathway, while her more vulnerable tearful response to the comments and the other parents' bids to connect (including the war stories) revealed the return to control of her ventral vagal system. In this state, she could take in the support offered by the other parents.

It's so helpful to remember that one real gift of the nervous system is that it has a preference for being in the ventral vagal state, the one that supports mutual regulation, care, and attachment. When another person in a ventral vagal state approaches, there is at least some probability that one can move out of sympathetic activation into a state that supports connection.

Through these examples of a student-teacher lunch regarding bullying, a principal's tea ritual approach to discipline, and a parent's metamorphosis, you can see how understanding some basic neuroscience can change how we approach staff, students, and parents. By now, it is probably clear that a felt sense of safety is developed through relationships with others.

Let's move from anecdotal individual examples and research-supported large-scale school district examples to practical tools you might be able to use in your school tomorrow. Each chapter will end with a Tools for School section, a description of techniques you can adopt and adapt to your classroom or school.

Tools for School

1. The 1 to 10 Scale. One method to help students monitor and keep their ventral vagal parasympathetic system online is through the use of a 1 to 10 scale. Simply put, the 1 on the scale refers to the state of a student who is relaxed and learning, and the 10 refers to the state of a student whose sympathetic nervous system or dorsal vagal system is active. When explaining the scale to students (and other educators), it is important to stress that it is normal to sometimes experience an 8, 9, or 10. When introducing the scale to your class (or an individual student), using yourself as an example by describing situations in which you have been at an 8, 9, or 10 can be helpful. The idea is to reinforce its normality. You can then expand the discussion to students by encouraging them to add examples of being at a 10 themselves. The emotions they discuss may be quite varied but will probably include anger and anxiety. You can explain that regardless of the emotion, at 10 no one can think logically and may do things they regret later.

Regardless of the emotion the students describe, once the extremes of the scale are understood, the emphasis changes to helping students realize when they are at a 5, the midpoint of the scale. At this point, they are experiencing an emotion, but still potentially have the capacity to calm it. It is important to discuss calming the emotional reaction and bringing the ventral vagal online in small steps. For example, "If you are starting to get frustrated with some work and recognize you are at a 6 on the scale, try to get your frustration down just a little, to a 5." You could go on to say: "When I begin to get frustrated, I sit back in the chair, take a few deep breaths, and try again. Or take a few breaths and try another problem and go back to the frustrating one later. Doing something like that helps me get my frustration down to a 5 or a 4. If I can solve a problem, my frustrations goes back to a 1. If I can't, I ask for help. Asking for help is a sign of courage and strength."

The emphasis should be on small steps because the activation of the sympathetic nervous system happens rapidly, but deactivation happens slowly. It is a neurological impossibility to immediately move from a high state of arousal (10) to a calm state (1). The goal is that once students realize they are at a 5 or higher, they will know they are still in a position to quiet the fight-flight-freeze response, either on their own or with support. Helping students get comfortable with asking for support when they need it is an important step, because our culture is so focused on autonomy and self-reliance. In truth, we are interdependent, and human brains were never intended to become completely self-regulating.

The next step is to help set concrete expectations for the alternatives available in the classroom to students in a state of activation. You will need several options. For example, sitting quietly at a desk and counting breaths (one to six in the inhale and one to seven on the exhale); other students will need physical movement, so a walk down the hall and back may be in order; others will need a distraction such as reading a book at their independent reading level. Still others will need to make calming contact with another person to borrow their ventral vagal state, so a visit to the guidance counselor may be appropriate. The options vary depending on the student, teacher, classroom, and school.

There are commercially available products that are similar to the 1 to 10 scale and include reproducible forms and posters for the classroom such as *The Incredible 5 Point Scale* by Keri Dunn Buron and Mitzi Curtis (2012).

2. Zones of Regulation. Another system to help students keep their ventral vagal parasympathetic system online is the zones of regulation curriculum. It is similar to the number scales, but uses colors. For example, the red zone is heightened sympathetic activation similar to numbers 8, 9, and 10 on the scale. The yellow zone describes heightened activation with some emotional control parallel to the middle numbers in the scale. The green zone is the calm state of alertness, coinciding with 1, 2, and 3 on the number scale. The author includes a blue zone to label lowered levels of alertness, such as boredom, sadness, or illness. The book was developed by Leah Kuypers (2011), an occupational therapist and autism specialist who has many reproducible sheets and a CD for teaching the zones. It is available at her website www.ZonesofRegulation.com or www.SocialThinking.com.

3. Professional Learning Community. At the school-wide level, initiating a professional learning community (DuFour 2004; Hord and Sommers 2008) is one possible structure that can develop trust among school staff. "The powerful collaboration that characterizes professional learning communities is a systematic process in which teachers work together to analyze and improve their classroom practice. Teachers work in teams, engaging in an ongoing cycle of questions that promote deep team learning. This process, in turn, leads to higher levels of student achievement" (DuFour 2004, p. 7). A well-functioning professional learning community can create a forum for teachers to dare greatly to be vulnerable and begin Brown's cycle of trust: "We need to feel trust to be vulnerable and we need to be vulnerable in order to trust" (2012, p. 47). The shared commitment of improving student learning unites the members of the community and can help teachers begin to make themselves vulnerable by sharing their successes and mistakes.

4. Appreciative Inquiry. Most managers who are trying to improve an organization (including schools) don't realize they have a bias: they look for weaknesses and problems within the organization and try to fix them. The assumption is that organizational improvement will happen automatically once the problems are solved. By contrast, appreciative inquiry, developed by David

Cooperrider, does not advocate that managers search for problems, but that they search for strengths. It answers the question: "What would happen . . . if we began all our work with the positive presumption that organizations, as centers of human relatedness, are alive with infinite constructive capacity?" (Cooperrider and Whitney 2005, p. 3). It strives for what management guru Peter Drucker suggested in an interview: "The task of organizational leadership is to create an alignment of strengths in a way that make a system's weaknesses irrelevant" (quoted in Cooperrider and Whitney 2005, p. 2). Organizational problems are not ignored, but are addressed only after an intense focus on individual and organizational strengths.

To explore appreciative inquiry further, a short overview is available in the book *Appreciative Inquiry: A Positive Revolution in Change* by David Cooperrider and Diana Whitney (2005) and at the website http://appreciativeinquiry.case.edu.

Chapter 3

Classroom Relationships: What's Love Got to Do with It?

THE RELATIONSHIPS UNFOLDING IN THE CLASSROOM HAVE EVERYTHING to do with whether our students learn, remember, and apply the information we are teaching. Yet the pressures educators feel to teach course content forces them to minimize the importance of relationships. This gives them the sense that there is no time to attend to relationships with anything other than consequences for difficult behavior. Interpersonal neurobiology teaches that if we only attend to relationships in a peripheral way, a large percentage of students will have unnecessary difficulty learning. The relational challenges they bring into the classroom create roadblocks in their brains (Cozolino 2013; Lieberman 2013; Siegel 2013).

As children move up through the grades and as the pressure to cover a certain amount of material in a fixed amount of time escalates, teachers can be increasingly pulled into left hemispheric modes of processing—the brain realm from which we teach facts, respond to the need for testing, focus only on judgments about how well students are doing academically, focus on teaching rather than student learning, and have strict standards for behavior without too much curiosity about why students misbehave. This makes attending to interpersonal relationships difficult because the neural circuitry of connection begins in the right hemisphere (and activates circuitry throughout the brain). When we are pulled into predominately left hemisphere mode processing, it is difficult to maintain a relational focus (McGilchrist 2009). The twist is that when we do focus on classroom relationships, students' brains become better equipped to take in information, which helps relieve the left-brain pressures through higher rates of quantifiable classroom success (Cozolino 2013; Lieberman 2013).

This chapter explores the neurobiology of attachment first, as a way to see our students (and ourselves) more clearly and to sense the importance of relationships to learning in the classroom. It then explores ways one can employ this understanding to create an environment in which students can learn more easily. The best part is that bringing the relational focus into the classroom helps our brains wire in patterns that support joy and well-being, increasing job satisfaction and decreasing burnout. With the higher test scores that are likely to result, there's something for everyone.

Let's begin by getting a sense of how students' earliest experiences make their way into the classroom. From the moment of conception, our survival depends on our relationship with our mother or caregiver. Her nutrition choices, alcohol consumption, cigarette smoking, and mental and physical health directly affect the developing structure of her child's body and brain. As early as three months after conception, her nervous system begins to shape ours (Field, Diego, and Hernandez-Reif 2006). By the time we are born, we have molded our heart rate (which tells us whether we are afraid or safe) and neurochemical make-up (which has much to do with mood and capacity for curiosity) to hers, building in the anticipation of whether the world will be welcoming and supportive of our efforts to learn or one in which we need to be so guarded that there isn't room for much new information (Field, Diego, and Hernandez-Reif 2006).

In the early months after birth, the core brain structure is being built. The billions of unconnected neurons in the infant brain are drawn into neural firing patterns, for better or worse, by the quality of the relationship with our mother, father, and others with whom we develop emotionally meaningful relationships. Their inner states, particularly the way they see us, begin to guide the aspects of ourselves we develop. If our parents are able to be warm, present, and curious about us, our inborn traits and natural aliveness can blossom. If instead our parents regard us as a disappointment or if they are too anxious or depressed to attend to our needs, we begin to develop a sense that there is something wrong with us. Because our brains are so unformed and so receptive at this early stage of life, these reflections get encoded at a foundational level and may well follow us throughout our lives. If our parents are anxious or distant, this will also leave its neural mark, guiding our expectations of how life can be for us as well as our behaviors in new situations, like school. Because one of our core drives is to attach, we become whomever we need to become to stay connected, because disconnection produces even more difficult challenges for our brains.

None of this information is intended to be critical of parents. Most parents do the best they can given their own wiring, their particular life histories, and the kind of support they received in their infancy and childhood. Instead, this information can help you get a sense of how influential the early days of our lives (and the lives of students) are in building the neural structures that then color our sense of safety or danger; our image of ourselves; whether we expect relationships to be supportive or hurtful; and the behaviors that arise from these deep patterns. When young, we develop perceptual filters through which we see all later experiences.

As a child leaves the relatively simple world of home and enters daycare or school, his or her brain is already shaped to some significant degree, but is still quite open to the influence of new relationships because of the ongoing development of neural structures. Teachers and peers then begin to help refine the already wired neural networks as well as create new ones in response to new situations. When a situation is novel, it supports increased opportunities for new wiring and also changes in old wiring, so in some ways, entering school gives a child another chance to develop some deep, long-lasting relational circuitry with the support of a conducive interpersonal environment.

Enduring Gratitude for a Fifth-Grade Teacher

A friend of mine, who grew up in highly traumatic abusive home circumstances, shared that her fifth-grade teacher was the first person who offered a safe, warm, and solid enough presence for her to attach, and the hope that this connection brought changed the course of her life. To help you get a sense of the power you have to positively influence your students through the forces of attachment and connection, here are some excerpts from an interview of my friend, more than fifty years after she was in fifth grade:

INTERVIEWEE: I had a fifth-grade teacher who I felt had warm wings around me the entire year, and I would say she was probably my first real attachment figure who helped me to survive. So there was that, along with some very big demands around academic performance as well. She really wanted us to do well, so I got the best of both worlds, really. She is a very significant figure in my internal life. I can see her and feel her presence this moment as I talk about her.

INTERVIEWER: And you are talking about her being an integrative force, really, of the [right and left] hemispheres.

INTERVIEWEE: Absolutely! And she was. I think she really knew which ones of us were suffering. I think she intuitively knew back in the day when there was really nothing about child abuse; no one was educated about that. She had intuitive awareness. It wasn't just me; there were three or four of us in that class that she just kind of helped the whole year.

INTERVIEWER: Well, I think we have to thank her right now. You have already thanked her, but we have to thank her.

INTERVIEWEE: I thank her every day. (Badenoch 2013)

It is vitally important to sense and remember the importance of what we bring relationally to our students. Their ability to learn depends on the level of support they receive, which in turn affects the status of their nervous systems and fosters neural integration of their brains (Cozolino 2013). In addition to cultivating a better environment for learning, these students will leave your classroom with a greater chance for fulfilling relationships (Lieberman 2013). Though the pressures in our school systems today may suggest otherwise, beginning with relationships and then moving to curriculum is the most efficient way to ensure our students' success, as we shall see.

The Power of One Sentence

Emphasis on relational dynamics is not just important for children. Let me share a story from graduate school. It was the late 1960s, and I had just graduated from college. I loved psychology and genuinely wanted to learn more. I was accepted into a master's program in psychology and was overjoyed to be studying further in the field I love. My new wife and I packed everything we owned into the back seat of our car and moved to the Midwest from New England. With eager anticipation, I sat in the front row of the first-day welcoming meeting for new psychology graduate students. The chairman of the department opened the meeting with one sentence: "Look to the left and now look to the right. Before the end of this year, one of you will be gone." I was stunned. It was a nightmare coming true. All I wanted was to learn more about a field that fascinated me. That one sentence changed the relationships among the students and between the students and the faculty. It changed my experience from eager anticipation of new learning to dread and competition. My nervous system accelerated, my ability to take in new information declined, and the mutual support that could foster learning moved rapidly out of reach. (The additional threat was that in those days, a man flunking out of graduate school would almost certainly be drafted to fight in Vietnam.)

As the semester moved ahead, nightmares plagued me. The chairman's "welcoming" speech set the tone for the web of relationships that developed in the school. Constant competition with other students prevented supportive friendships from developing; grades for tests and reports were posted in public so we could all see who was not going make it; and no one could to be trusted. Essential reference books disappeared from the library and reappeared after papers were turned in. I learned to go directly to the library from class to get my hands on copies of the books and articles I needed. I honed the sordid art of competitive studying. I survived the first semester living in a world of distrust; my initial enthusiasm for new learning now seemed naive. That single sentence, uttered by the department chairman, who lacked understanding of the relational water in which all students and teachers swim, caused unnecessary conflict and pain, and, I argue, poorer learning outcomes for the graduate students.

The second semester, I walked into an introduction to industrial psychology class and met a professor who changed my life. He assigned a weekly ten-page paper. I turned in my first assignment, reasonably confident. It was returned covered in red ink, including the dreaded phrase: "See me." I waited outside his office, worried that this was the beginning of the end of my career in psychology. He welcomed me with one sentence: "Your ideas are wonderful; let me help you express them better." He then carefully and compassionately went through my grammatical mistakes and made my awkward prose flow smoothly. I was amazed how moving words around a page could make both my thinking and my writing clearer. In spite of the negative relational culture set by the chairman's words, this professor's one sentence began a positive student–teacher relationship. His office became a safe island within an ocean of negativity. I learned not just industrial psychology but what it is to be a professional and how to reach for excellence.

My experience from more than forty years ago demonstrates how important it is for educators to understand and harness the invisible web of human relationships that creates the context for student learning. As was the case with my professors, some teachers have a deep understanding of the relational context of learning and some do not. To my knowledge, the relational contexts of learning are not explicitly taught in most teacher training programs, yet some teachers have learned it. How is this possible? Most likely, they learned through the secure connection they made with their parents. Let's delve into the research so we can understand our own teaching as well as the connections we have with our students, other teachers, and administrators. This knowledge will help us begin to make the invisible web of relationships visible.

The Social Brain and Human Attachment

The field of study devoted to how humans connect with each other is summarized by the term *attachment*. The inception of research and theorizing about human attachment is generally attributed to British psychiatrist John Bowlby, whose research in the 1950s and 1960s culminated in his three-volume *Attachment and Loss* (1973, 1980, 1982). He carefully studied the intricacies of the mother–infant dyad, and noticed that not all attachments were the same, even at this early stage of life. In some dyads, there was an easy flow between the mother and child; others were fraught with the unpredictability of overwhelming attention followed by dismissiveness; and for others there appeared to be almost no connection at all.

Mary Ainsworth and colleagues (1978) expanded Bowlby's work by studying infant attachment, using careful home and laboratory observation along with an interesting methodology she called the strange situation. While there are many variations of this process, the core aspects of her protocol included three phases of relationship: first, one-year-olds were observed with their mothers in an unfamiliar playroom; next, their mothers left them in the playroom with a stranger (one of Ainsworth's research associates); and finally, the mothers returned. Again, each of the mothers had also been observed at home.

Ainsworth found that approximately 65 percent of the infants used their mothers as a base for exploring the unfamiliar playroom, were distressed when their mothers left, and were comforted when they returned. They were easily soothed and eager to return to play after the reunion. She called this group of infants *securely attached*.

Another group of infants did not use their mother as a base to explore the unfamiliar playroom, showed little distress when their mothers left, did not seek comfort from their mothers when they returned, and showed no more preference for their mother than for the stranger. Because of their parents' disconnection from emotional life at home, these little ones had not learned to value relationships, but instead were used to going it alone. She called them *avoidantly attached.*

A third group of infants was preoccupied with their mothers' availability when they were present in the unfamiliar playroom; showed distress when their mothers left; and met their return not with warmth, but with anger, clinging, or resistance; and were not easily comforted. They had learned that their mothers' ability to be present with them was unpredictable, so the young ones needed to watch them closely for signs they were going to become unavailable again. She called this group *anxiously/ambivalently attached.*

Years later, Mary Main and Erik Hesse (1990) noticed that some of the infants' responses did not fit comfortably into any of Ainsworth's three attachment styles. These infants showed a variety of intense, unexpected responses to their mothers' return to the playroom, including freezing with a dazed expression on their faces, approaching their mothers with their heads averted, and simultaneously showing other contradictory behaviors. They called this fourth attachment style *disorganized/disoriented attachment*. This occurred when the infant saw the caregiver as frightened or frightening. If the child went toward the parent, the person to whom she would turn for comfort, she experienced fear, and if she turned away, the fear that comes from being alone was present. Main and Hesse saw this as "fear without resolution." The infant might be frightened of the parent because of painful experiences in that relationship (for example, physical abuse or repeated anger), or the caretaker might be afraid when having contact with the infant, something the child's system internalizes as equally frightening. The disorganized/disoriented group of infants represented the smallest percentage but exhibited the highest level of relational difficulty and mental illness later in life.

Avoidant, anxious/ambivalent, and disorganized/disoriented attachment are often collectively referred to as *insecure attachment*. One of the most astounding parts of this is that we have a hard-wired, recognizable attachment pattern in place by one year of age. Humans are still carrying remnants of these initial attachments, regardless of our intellectual ability or age. This means that these attachment styles add to the complex web of relational expectations coloring every interaction in our classrooms.

Attachment Style Is Not a Passing Fashion

The term *attachment style* may be confusing because the word style makes it seem that it is something that can be easily changed—like clothing styles that change from year to year. This is not accurate. To help us understand the pervasive influence of the attachment styles established in early life, let's briefly revisit the topic of the early formation of neural networks in the brain.

For infants, the relationship with the primary caregiver causes some networks of neurons to connect, while other networks do not have a chance to grow in the fragile, developing infant brain. This creates deeply established core brain structures, carrying patterns of relationship that persist into adulthood. As the infant matures, brain cells that are not used will eventually die in a process called *pruning*. There is a good deal of neural pruning around the age of three and again during the teenage years (Siegel 1999, 2012a, 2013). The death of brain cells during stages of development is sometimes misunderstood as a negative event, but this is not accurate. The pruning process makes the child's brain more efficient because it eliminates the unused networks and leaves in place those networks that have been strengthened by repeated experience. For example, if the infant has a caring, consistent parent, then a complex network of neurons will fire every time the child is cared for. When the child explores an unfamiliar room, becomes anxious, returns to the caring parent, and is calmed, a complex network of cells fires. Over and over, the infant has a similar experience, and the same neurons fire in roughly the same pattern. As we said before, what fires together wires together.

It is important to remember that neurons fire for every kind of experience, positive and negative. Neurons will connect if the parent scolds the child for returning when he or she is anxious or if the parent comforts the child.

Attachment Styles in the Classroom

As infants mature into toddlers, preschoolers, and eventually school aged students, the early neural networks become hard-wired templates that tell them what to anticipate in relationships and then guide behaviors that have a tendency to make these predictions come true. If I have an expectation that people will not help me, I may either avoid contact or lash out in defense when I need help. These actions make it much less likely that people will be drawn to help me. If I grew up in a household where there was a lot of criticism and I adapted by becoming angry, I will come into my first-grade classroom expecting to hear criticism and probably lash out or engage in other behaviors that will draw more criticism toward me. These cycles become self-reinforcing unless a teacher with wise eyes sees what's going on and provides a different and disconfirming experience.

Attachment templates are established before the cortex has developed enough to assess them in any meaningful way. These foundational neural networks have established how we perceive the world before we have the neural equipment to consider other options (Cozolino 2013). Attachment styles, developed in infancy, become nonconscious filters for the relational world, and this includes the child's perception of you, the educator (Badenoch2008, 2011; Siegel 2012a). The patterns developed in infancy mean that students will expect certain behaviors from the teacher.

Securely attached students with caring and consistent parents have developed the neural networks to expect helpful attention from adults, and they have developed ways to effectively elicit that help. When they experience something that raises their anxiety, such as a final exam in your classroom, they "know" that the anxiety can be lessened by contact with trusted adults, just like all those times during their earlier childhood.

Insecurely attached students are likely to respond in one of the following ways:

- Avoidantly attached students will not care much about you and may have a tendency to become angry, but might be interested in the academic content of the class. If academics were the only concern, some of the avoidantly attached children might be ideal students.
- Anxiously/ambivalently attached students likely will be overly concerned about you, may seem clingy, and might worry if you can be trusted.
- Students with disorganized attachments will react to you in various confused and emotionally dysregulated ways, particularly when they are under stress.

Generally, those students with insecure attachments are more vulnerable to anxiety and may perform poorly on tests, not because they don't know the material but because their insecure attachment makes them less able to manage the anxiety that is a normal part of final exams, school, and life in general. Insecurely attached children will make up a third to half of your students depending on the community from which your school draws (Bergin and Bergin 2009). The upset within them may negatively influence other students. Intentionally creating a classroom culture supporting a felt sense of safety may help repair the relational circuits for insecurely attached students and lessen their effect on other students.

Attachment Style and Emotional Control

Early attachment also influences our capacity to regulate our emotions. New learning elicits an emotional response, be it excitement, fear, frustration, or a combination of these and more. Securely attached students are likely to find the normal emotions elicited by new learning relatively easy to navigate, whereas insecurely attached students tend to have a more difficult time regulating emotional responses to new learning. Avoidantly attached students tend to react with anger when they come across a challenge; anxiously attached students react with fear and worry; and those struggling with disorganized attachment respond with a contradictory pattern of intense emotions that are as bewildering to them as they are to you. In general, this difficulty with emotional regulation makes learning difficult for insecurely attached students regardless of their intellectual level (Bergin and Bergin 2009; Cozolino 2013).

In addition to relational expectations and emotional regulation, attachment style is associated with vulnerability to the behavioral difficulties we see in our students: "In general, ambivalent attachment predicts vulnerability for anxiety problems, whereas avoidant attachments and disorganized attachment predispose an individual to develop conduct problems" (Siegel 2012a, p. 114). Certain students may come to mind as you absorb this way of perceiving attachment patterns. As you allow this information to influence your perception of your students, you may gradually notice a decrease in negative judgments, a rise in empathy, and an urge to meet them on this new ground of understanding.

High Expectations Alone Don't Lead to High Performance

Most teachers believe that maintaining high expectations for students leads to a high level of performance. It is a simple, appealing concept, and any reference to maintaining high expectations is usually met with knowing smiles and nods. This is also the implied and sometimes overt belief behind state and national educational policies and mandatory academic testing. The seeds of this nearly universal belief were planted by a study from the 1960s titled "Pygmalion in the Classroom" (Rosenthal and Jacobs 1968). Briefly, in the original study, teachers in an elementary school were given a list of students whose scores on the "Harvard Test of Inflected Acquisition" indicated that they would soon have an intellectual growth spurt. Later in the school year, all of the children were tested again; the academic achievement and IQ scores of the children on the growth spurt list had risen significantly more than the other students in the class. However, the Harvard Test of Inflected Acquisition was not a real test and the "growth spurt" children had been chosen randomly. The rise in scores of the randomly chosen students was attributed to the teacher's higher expectations of them. The publicity surrounding this study and other studies with similar results has turned the results into a firm belief that students will always rise to the high expectations of teachers. As discussed, students are definitely tuned in at a very deep level to how we see them. At the same time, research on the neurobiology of relationships shows that maintaining high expectations is only one part of a more complex story.

Securely attached students with adequate cognitive skills have likely developed the neural circuitry to be able to meet high standards set by educators because they have the capacity to persevere in the face of the frustration, worry, and excitement they experience in the process of meeting the expectations. However, insecurely attached students may well have difficulty managing the emotions—positive and negative—that arise from the school experience. Insecurely attached students need both high standards and the disconfirming experience of a consistently emotionally available and supportive teacher to succeed. As my friend's fifth-grade experience attests, the combination of the two can be life changing.

You may remember we talked about disconfirming experiences in Chapter 2. These occur when a person's neural network actively anticipates a particular outcome of an event, but something different occurs. For example, if an insecurely attached student's neural network anticipates a teacher's scorn when the student fails a test, but instead finds that the teacher is warm, accepting, and helpful. The combination of an activated neural network and an unanticipated helpful response creates a disconfirming experience that may lead to weakening the old network and developing a new, more positive network (Ecker et al. 2012). A series of disconfirming experiences may contribute to a change in a student's attachment style. It is similar to the experience Ms. Atkins designed for Jimmy and Tucker in Chapter 2. However, an attachment style is more foundational and likely more difficult to change than Tucker's negative school neural network.

High Expectations and Emotional Support Are Life Changing

Elisha is a fifty-eight-year-old woman with an anxious attachment style that is probably the result of parental abuse. She remembered the positive effect of her relationship with her high school English teacher, Mr. Blanden, in an interview with me:

> He helped me change how I viewed the world and how I viewed myself by who he was. He made my view of the world more expansive about things larger than [my hometown]. There were beautiful things in literature and the arts and music and culture and plays. He made them fascinating. Made me challenge myself and think about the fact that I was more than I thought I was. He had high expectations for himself and for us. I can see him as clearly today as if it was all those years ago. He was tall, his posture was confident, his hair was thick salt and pepper brushed back in an Einsteinish way, but neater—a red bow tie and a suit with a formal brief case. He would walk into class with a lot of energy and stand behind the podium and talk animatedly. He was so excited about what he was talking about you felt the excitement. When he looked at you, you felt he was looking at just you and not the whole class. I was shy, so I talked to no one, but I felt connected to him not like in a friend way, but a connection with his passion and energy about the subject matter. It was the hardest class I had ever taken. And he had the clear expectation you could step up to the plate. He was a unique individual. He was like a mirror; if he could do that, why couldn't I? He is the only person I ever think about when I think about teachers, throughout my whole experience in school. (Olson, 2006, p. 1)

Elisha's detailed, vivid memory of Mr. Blanden more than forty years after she was in his class shows the power of the high academic standards and the disconfirming experiences he created for this shy student. (In another part of our interview, she described being so shy she often hid in the girl's room during lunch period rather than face the school cafeteria.) Mr. Blanden created multiple disconfirming experiences during the school year through a combination of his high expectations ("it was the hardest class I have ever taken") and his disconfirming expectation that "you could step up to the plate." Although Elisha's abuse continued in high school (she revealed it to no one), these multiple disconfirming experiences changed her view of herself ("He helped me change how I viewed the world and how I viewed myself"), first within the context of his classroom and later within the larger context of schooling. She credits her experience in his classroom as one of the main reasons she was able to go on to earn bachelor's, master's, and doctoral degrees, something unthinkable prior to his class. All her years of education make her last sentence especially powerful: "He is the only person I ever think about when I think about teachers."

The disconfirming experiences Mr. Blanden created were more subtle, complex, and effective than simply having high expectations for his students. His capacity for demonstrating that he saw students as competent enough to rise to his high standards may have helped them bring those aspects of themselves into existence. Elisha's interview suggests that her experience in Mr. Blanden's class brought to life her confidence in her own academic ability. His dual expectation of high performance and confidence in each student's ability to meet high standards was further confirmed by my interviews with five of his other students more than forty years after they had graduated from high school (Olson 2006).

Attachment research can seem like it describes a mysterious process, and Elisha's experience can, on the surface, be seen as the result of the unique gifts of a charismatic teacher. In the past, it may have been possible to take either of the extreme views of dismissing or marveling at Elisha's experience, but these are no longer the only options. The expanding research base in the fields of neuroscience and human relationships means that educators don't have to be narrowly focused on students' cognitive learning, knowledge, and skill acquisition. We can include their emotional and relational learning as well. This is true even if we were to adopt the view that tested academic achievement is the only measure of school success, because emotional and relational learning support academic learning. Excellent, experienced teachers have known this for a long time; research in neuroscience and positive psychology now gives us the solid scientific underpinnings for what they know intuitively.

Secure Attachment Is Not Dependence

It is important to understand the difference between secure attachment and dependency. If we focus only on the child–adult pair, we see a distressed child asking for nurturing and the adult responding. In that moment, the child is dependent on the adult for comfort and regulation—by which I mean soothing of emotional need until anxiety subsides. During the moment of soothing, an uneducated eye might see "a clinging crybaby." But as time goes by, the securely attached child learns that the adult can be depended on for soothing when her anxiety rises or when she shows subtler forms of distress. In short, the securely attached child learns she does not have to be a clinging crybaby to have her needs met (Sroufe and Siegel 2011). If we teachers are able to notice the first signs of anxiety or frustration and soothe the child quickly, we can reduce excessive crying and neediness later. Responding rapidly to small cues teaches the child that the teacher can be relied on, and so the child has less need to loudly express his or her distress. For example, we can initiate this process by approaching a child who is quietly standing on the periphery of the playground on the first days of school or quickly offering help to a student at the first signs of frustrated erasing on a paper. This is especially important information if you are a day care, preschool, or kindergarten teacher. If attachment needs were not met earlier in life, the neural circuitry in a student of any age may be still similar to that of a young child, and so are their legitimate needs for soothing.

The difference between dependency and secure attachment is clearly seen as we expand our attention to the child's exploration of new learning. The securely attached child, reassured by the presence of a responsive teacher, quickly resumes exploration and new learning once anxiety has been soothed. The availability of reassurance from a balanced adult and the child's capacity for curiosity and independent investigation are strongly linked. Your presence as a nurturing adult in the classroom may often be enough to provide the support to turn on curiosity and dramatically increase the effectiveness of your teaching (Panksepp and Biven 2012).

On the other hand, dependent, anxiously attached children spend much of their emotional energy and time focused on the adult, trying to get the nurturing they need. The inconsistent parenting they received means they are never really sure if the adult will be physically and emotionally present when needed. They must spend their energy checking and rechecking the adult, and when the adult is able to be present, the child tries to get as much nurturing as possible because he doesn't know when it might be available again. This does not leave the child with energy for exploration and new learning. As noted in Chapter 2, when the nervous system is elevated into sympathetic activation, attention narrows to just information related to the perceived threat—in this case, the threat that the adult might become unavailable at any moment.

This situation can be difficult for a teacher. From my time as a preschool teacher, I remember a young boy who demanded to sit on my lap at morning circle and was "attached to my hip" throughout the day. It was difficult and frustrating to try to attend to other children when one was always present. It can be even more difficult when there are several who compete for this connection at the same time. Because I was that rarity—a man in a preschool—I often had this experience. I wish I knew then that these children were probably insecurely attached and not trying to make my work impossible. As I learned in that preschool classroom, ignoring or telling these children to do something else only increases their clinging behavior, whereas offering them the disconfirming experience of a reassuring supportive presence gradually increases secure attachment and lessens the challenging clinging behavior.

Looking at both patterns, paradoxically we see that the adults in the life of a securely attached child have accepted the youngster's periods of natural dependency as a supportive stepping stone toward courage, curiosity, and exploration. The parent of the insecurely attached child has not been able to meet the child's need for dependency before the child starts school, so this child continues to seek a secure footing from teachers before being able to move outward. High expectations alone will not help these students. Teachers with equal parts high expectations and supportive, reassuring presence will foster independent learning by offering a reliable connection for students when they need it.

I have observed the natural blend of high expectations and a reassuring presence with many teachers, and the memory of a calculus teacher stands out in my mind. She was a consistently encouraging cheerleader for her students. When they became frustrated with their mistakes, she reassured them by explaining that mistakes were necessary for learning to occur. "You are learning something new. Of course you make mistakes. If you never made any mistakes that would be my mistake, because I would be giving you work that was too easy! This work is hard! You can do it!" Obviously, her students loved her, even when they feared the subject she taught.

Neuroplasticity and Attachment Style

Fortunately, the human brain is continuously changing in response to new experiences—due to neuroplasticity, as already mentioned. Early experiences do not doom students to a life of hard-wired, predetermined reactions (Doidge 2007). It is possible to modify attachment styles. Individuals who have had early childhood experiences and developed circuits for insecure attachment can develop earned secure attachments through connections with friends, romantic partners, therapists, teachers, and others (Siegel 1999, 2012a).

Research has shown that teachers and schools can have a positive effect on students' attachment style. In their study of school dropouts who returned to complete a General Education Development (GED) program while still teenagers, Reio, Marcus, and Sanders-Reio found "the magnitude of the [statistical] effect demonstrated by these findings suggests that student–student friendships, student–instructor relationships, and attachment each have a pronounced effect on General Educational Development program completion" (2009, p. 66). In other words, the quality of the relationships students developed within the classroom had a positive effect on the probability that high school dropouts would persevere and complete their GED.

The 50 to 65 percent of your students who have a securely attached style (Bergin and Bergin 2009) will generally meet you with a positive attitude, and you will probably respond to their positivity in kind. Teachers tend to treat securely attached students warmly and hold them to high standards of behavior. The 35 to 50 percent of your students who are insecurely attached will be the most difficult to teach, and doing nothing to address their needs ensures that their pattern of difficult behaviors will continue.

Although it may lead to short-term compliance, it is unlikely that reward or punishment will change someone's attachment style. Actually, use of external incentives alone runs the risk of strengthening the insecure attachment circuits. Consider, for example, anxiously attached students. Depending on a particular child's temperament and a host of other factors, focusing on reward and punishment can lead to clinging or angry outbursts. Anxiously attached students' early experiences of caregivers who are sometimes available and sometimes not keeps them in an anxious state and focused on whether you will be available. Some of their classroom misbehavior, no matter how awkward or annoying to you, is simply an attempt to connect. Punishing or ignoring the behavior may stop it for a short period, but it reinforces a core belief that their distress doesn't matter, leading to more distress, more misbehavior, less learning, and frustration for you. To begin to transform the root of the problem and break the repetitive misbehavior cycle, you need to look underneath the behavior to the core need that is being expressed and deal with it by taking corrective preventive relational action to create a solid learning environment.

You might get the impression that one needs to be a psychologist or psychiatrist to deal with these core beliefs, but this is not the case. Simple techniques can have profound effects, and you can build them into the culture of your classroom. They provide disconfirming experiences for insecurely attached students, setting the stage for long-term positive change in their attachment styles and, in turn, their ability to learn.

The Teacher as an Emotional Base

Our discussion has so far focused on the development of secure and insecure attachments in infants and young children. As these youngsters enter school, their attachment circuits are not fully formed, so preschool, kindergarten, and first grade teachers have an important opportunity to positively affect attachment circuits in a direct way (Bergin and Bergin 2009). However, all teachers, regardless of the age of the students, can become an emotional base for them, fostering a secure attachment that can even generalize to the whole classroom and school. Secure attachment with a teacher supports a zest for learning just like it supports exploration in the infant. For example, Gregory and Weinstein (2004) found that adolescent perceptions of their connection with their teachers were the strongest in-school predictor of academic growth in mathematics from eighth to twelfth grade. An interesting additional finding was that the combination of a positive teacher–student connection with high standards predicted the most growth in students from lower socioeconomic backgrounds.

Securely attached relationships take time to develop, and some schools unwittingly make this difficult. For example, it is more likely a securely attached relationship will begin to develop in an elementary school simply because the teacher spends most of the day with the same students. In this model, developing secure attachment is interrupted at the end of the school year because the students move on to another grade, making access to the previous year's teacher difficult. In high school and many middle schools, the opposite occurs. Students have five to seven classes a day, are taught by different teachers, and do not spend much time with a single teacher, but these students are more likely to have contact with a single teacher over several years. An aspect of the secure attachment relationship is that the child must have the opportunity to return to the attachment figure when anxious, be soothed, regain feelings of security, and venture out again.

Long-term securely attached student–teacher relationships support learning. Perhaps schools can make it possible for students to meet with previous teachers within the school structure (e.g., during study hall, between classes, or during school activities). Simply spending time with students is the first necessary (but not sufficient) step to create securely attached relationships with students.

Building Attachment into School Structure

Some middle and high schools have developed advisory, mentoring, or guidance counseling programs. These kinds of programs assign students to specific teacher/counselor advisers based on shared interests or a positive connection, rather than by random assignment, such as alphabetically by last name. The adviser–student pair continues to communicate regularly for several years. A growing body of research shows that building such organizational structures increases student engagement and academic performance (Klemm and Connell 2004).

There are several methods to develop an advisory program, and they have a few common factors. First is a formal method for teachers and students to be assigned to each other. Allowing the students to have an active voice in choosing their adviser is important because it helps them buy in to the process and supports the adviser–student connection. The second and most important aspect of the program is time. Scheduling time for the advisers and students to meet is crucial. Rather than being an add-on that takes time away from learning course content, a well-functioning program creates an opportunity for authentic connections and secure attachments to grow between students and teachers, increasing student engagement and improving academic performance (Klemm and Connell 2004). Third, the adviser–student connection is best if it continues for several years; for example, if high school first-years maintain the same advisers for all four years. Finally, as with any program, there are a few minor problems to be solved: some teachers have a large number of students ask to be advisees and some do not. Simply having a fixed and equal number of advisee slots for each teacher solves the problem as well as explaining to students that they may not get their first choice of adviser. Having a formal mechanism to change advisers solves the problem of the occasional mismatched student–adviser pairs. Once in place, the adviser–student time can be a formal discussion of a pertinent topic with all of a teacher's advisees in one room (e.g., a discussion of life after high school, planning an event, or choosing classes for the next year) or informal time (e.g., a shared lunch with one student). Formal or informal, group or individual, the goal for the teacher is to foster secure attachment by being a consistent, supportive, stable adult in a student's life.

It is interesting to note that students within the special education system may have more opportunities for developing a secure attachment to teachers. Many times students who receive help for learning, emotional, or other challenges have the same teacher or specialist (speech therapist, occupational therapist, etc.) for many years, making a secure attachment more likely. This can be taken as an opportunity to encourage secure attachment for students who have learning challenges.

Advisers and Attachment: A Case Study

Al arrived at the school shy, depressed, and surly. He was often seen sitting in the hall, staring at the floor. Even when approached pleasantly by a staff member, he sat there, said nothing, and, except for an occasional nod, was uncommunicative. He lived alone with his father, who was rightfully worried about his son. I remember from years ago the anxious and confrontational tone of his father's questions in a parent meeting.

"Are you going to tell me what I want to hear and then do nothing for my son, or are you really going to help him?"

"Will my son ever snap out of it—feel better about himself and do okay in the world?"

"Will he ever stop sitting alone in his room playing video games?"

As the meeting wound down, his father gave us the hint we needed: "Al is interested in playing the drums, but no way am I going to put up with all that noise. Besides, he never follows through on anything."

The next day, Al begrudgingly agreed to change a study hall period into a music class. The music teacher, who had a degree in business and was lead singer in a well-known local rock band, felt that Al was reaching out to him. He helped expand Al's connection with other faculty by facilitating an adviser assignment with another teacher that lasted for the next four years. During that time, the music teacher taught Al drums and guitar. Al performed for the parents and others at one of the school's open mic nights. He calmed his anxiety before the performances by discussing it with his adviser and seeing the anxiety of the other students helped him normalize it. He joined a rock group made up of several older students at the school. Al, his adviser, and his music teacher were able to convince his dad to buy a drum set as long as he agreed to a specific schedule of practicing at home. Under the tutelage of the music teacher, the rock group turned into a lesson in business as well as music. Like most budding teenage musicians, they had a dream of recording their own CD and selling it, so the music teacher challenged them to earn the money needed to book a recording studio. Al suggested they make toasted cheese sandwiches at lunch and sell them to the students and staff. It was a huge success, and Al was on hand every lunch, using his developing sales skills to sell sandwiches that he cleverly named after popular staff members.

The metamorphosis from an uncommunicative boy to a young man using sales tactics to move sandwiches was heart-warming.

At the core of his change were the secure relationships with his adviser and music teacher. When Al became depressed or anxious, he talked with both. They had frequent contact with his father. Through the whole process, both teachers were a presence in his life: commenting on and giving him the opportunity to demonstrate his strengths, carefully raising the expectations after Al met lower ones, and consoling him when necessary. It was a long process, but Al's secure attachment to his adviser, a music teacher, made it possible for him to develop positive connections with other school staff and even the school itself. From an anxious, depressed, and uncommunicative boy, Al grew into a seller of sandwiches and then to a mature school leader who helped other students. His interests broadened when he realized that he loved music but was unlikely to become a rock star. He contemplated taking psychology or business classes in college. During his senior year, he enrolled in a class at the local community college and did well. He successfully graduated from high school a confident, insightful young man.

That is not the end of this story of teachers, advisers and attachment. After Al did well in the first semester of college as a part-time working student, he felt ready to take on a full-time college schedule for the second semester, which turned out to be a mistake. His school performance crashed. During the crisis, he called his old adviser and music teacher for help. Together they came up with a plan and approached Al's father. Al quit his job, reduced his class load back to part-time. He was appreciative for the help and offered to give back to the school by teaching younger students the drums. He is doing well in college and maintains his contact with his adviser, music teacher and the school in general.

Al's story is not one of an unusual student or even exceptional teachers. It is a story of an average student and his teachers who developed secure attachments because the structure of the school encouraged it.

Teacher Attachment Styles

Although it is important to consider the effect that your students' attachment styles have on their reactions to you, it is equally important to consider the effect that your own attachment style has on your reactions to your students. Research on adult attachment style is complex, because adults have had more attachment experiences with friends, teachers, mentors, and romantic partners. The vocabulary is also slightly different from the vocabulary of childhood attachment styles, although it continues to follow the secure/insecure polarity. The types of insecure adult attachment style include (Siegel, 2012a):

- Dismissive: people who are comfortable without close emotional relationships, value independence, and are uncomfortable with being dependent or being depended on by others.
- Preoccupied: People who want to be continually emotionally intimate with others and are uncomfortable if they are not in a close relationship. They tend to worry that others do not value them as much as they value others.
- Fearful/ambivalent: People who want a close relationship but cannot trust or feel dependent on others. They worry they will be hurt if they get too close.

All teachers have experienced an immediate negative reaction to a student. In some instances, the response may be caused when the adult attachment style conflicts with the student's attachment style. For example, consider a teacher with a dismissive attachment style, who avoids close relationships, and how she might react to a student with an anxious/ambivalent attachment style who tends to cling to the teacher. Or the opposite: a teacher with a preoccupied attachment style who may worry if a student doesn't like him, and how he might react to a student with an avoidant attachment style who does not care much about the relationship. There is potential to create repetitive negative interpersonal cycles that become the defining element of particular teacher–student relationships—a miserable situation for both involved. Thankfully these negative cycles can be interrupted and minimized when we are able to identify the attachment styles in both students and ourselves.

It is important enough to reiterate that the automatic nonconscious responses students have to us may have little to do with the teacher personally or with teaching style. Likewise, the instant nonconscious responses to students are influenced by attachment style and may have little to do with the students. As noted earlier, nonconscious perceptions occur through brain circuits in the limbic region and right hemisphere. These circuits react in tiny fractions of a second, much more quickly than circuits involving conscious thought. The whole process begins in less than a second of the first day of class, and without some attention from the teachers, can define the year with students whose attachment styles are different and conflict. There is no need to allow this to happen! Self-exploration can help one interrupt the process before the immediate reaction defines the relationship. It becomes a problem to be solved when one realizes that the cause of the friction with a student does not reside within the student nor does it reside within the teacher but is caused by the mismatch itself. Coming to grips with these inevitable mismatches can be helped with discussions with trusted colleagues in forums like the professional learning communities outlined in Chapter 2 or the circles of trust described in Tools for School at the close of this chapter.

There is some cautionary research by Philip Riley (2011) who explored the effect of attachment style on teachers who became verbally or physically aggressive in the classroom. He hypothesized and found that some people choose the teaching profession not just for the love of teaching children but because they are unconsciously searching for an experience that will repair their own difficult childhood attachment experiences. This does not always work out well. Teachers who unintentionally find themselves in this predicament can experience strong emotions in the classroom without realizing why or how to modulate them. Riley developed a structured series of interviews to help teachers understand their emotional reactions, learn how to control them, and possibly begin to develop the earned secure attachment for which they are striving outside of the classroom (Riley, 2011).

It is important to note that one doesn't have to be perfect in relationships with students. Disruptions are normal in any relationship—psychologists call them a "rupture" in the relationship. Ruptures will occur—it is what you do next that is important. There is research showing that relationships become stronger after a rupture has occurred and been repaired (Siegel 1999, 2012a). It can give hope to know that you only need to meet your students with accurate empathy 33 percent of the time as long as it is followed by a repair (Tronick 2007, Siegel 2013).

What's Love Got to Do with It? Everything!

How many times when you are reflecting on your students have you publicly or privately said, "I love these kids?"

It's 5:30 pm and I'm leaving the school building after a tense meeting about a student involving parents, lawyers, the principal, and others. I'm tired and feel beat up. As I walk down the dimmed hallway, I notice the lights on in a classroom and Ms. Atkins sitting at her desk, staring into space. Those who frequent schools well after the students have left know the experience of being so exhausted from teaching all day that you don't even have the energy to pack up your things and go home. It can feel like being caught in purgatory between the two worlds of school and home.

I go into her classroom, disrupting her reverie, and with no acknowledgment she begins: "I just love these guys, but Jimmy is driving me nuts. I can't figure out what is going on with him now. Last week he was doing really well, especially after we sorted out that mess with Tucker. Tucker is doing a lot better, by the way. He is realizing that being a 'motor head' and an outdoors kind of guy can be a good thing. Remember the car trouble I had the other day? I started talking about it with the kids; after all, I had to explain why I was late. He actually was right about what was wrong with my car. I told him I should have had him fix it rather than pay for it to get towed. He is really growing on me. But Jimmy, why does he sometimes speak up and other times seem like a scared rabbit?" She smiled, her monologue continuing as if I were not there. "Maybe he is gradually learning, sometimes he can overcome his fear and sometimes he can't. I sure love that kid. He has such potential if he can only conquer his fears. I'm even learning to really like Tucker. They are both coming along. I can just feel it." Her speaking seemed to energize her a bit. She looked up and, as if seeing me for the first time, asked suddenly, "What time is it? I guess I'll pack up and head home. People just don't know how much we work from the heart. Sometimes I think I love these kids more than their parents do. Is that weird?"

Many of us have had similar conversations and feelings. Although the word love is not an unusual word for teachers to use in regard to their students, its use has always sparked my curiosity. The teacher–student relationship does not fit into the usual categories; it's not the same as the parent–child relationship and certainly not a committed long-term romantic relationship. Some interesting research by Barbara Fredrickson (2012), a well-known researcher in the field of positive emotions, helps explain the neuroscience behind this common phenomenon. In her book *Love 2.0*, she states that love is an emotion and goes on to say, "Put it in a nutshell, love is the momentary upwelling of three tightly interwoven events: first, a sharing of one or more positive emotions between you and another; second, a synchrony between your and the other person's biochemistry and behaviors; and third, a reflected motive to invest in each other's well-being that brings mutual care. My short hand for this trio is *positivity resonance*" (2012, p. 17, emphasis in original).

Fredrickson outlines two preconditions for love: (1) safety and (2) face-to-face contact. Remember, safety first for learning; Fredrickson says this mantra is true for love as well. Her second precondition for love is face-to-face contact with the individual(s). Noting that one can "stay connected" through social media, telephone, and email, she says that the mind/brain/body has evolved to require face-to-face contact for love, or in her terms, positivity resonance, to bloom. Jing et al. (2012) and Yun (2013) elaborate on Fredrickson's findings with brain scans that show neural synchrony that occurs during face-to-face contact that does not occur when there is contact using electronic media.

Fredrickson further elaborates on the first of her trio of positivity resonance, "the synchrony that occurs during moments of shared positive emotions," by describing the imaging research of Uri Hasson and colleagues (2010; Stephens, Silbert and Hasson 2010). Hasson's team mapped the brain of one person telling an engaging story about her life, then had others listen to it while their brains were being mapped. He quizzed the listeners on the details of the story to measure how much they were engaged with it. Some of the same brain regions were active in both listener and storyteller. This was unexpected, because the act of telling a story is quite different from the act of listening to it. As you might expect, most of the listeners' brains reacted about one or two seconds behind the storyteller because it took them time to process the story. A more interesting part of the research was that the more the listeners were engaged with the story, the closer in time their brains responded to the storyteller's brain. The real surprise in the research was that the brains of the people who paid the closest attention, who hung on every word and were able to recall even extraneous bits of the story, actually *anticipated the storyteller's brain.* Hasson calls this *brain coupling* and claims that true communication may be a single act performed by two coupled brains (Hasson et al. 2011; Fredrickson 2012;). If you delve a bit deeper into the research, you will find that one brain area that showed synchrony between the storyteller and the listener was the insula. This area of the brain is believed to be the link between the body and the brain, and it takes part in conscious thought (Siegel 2012a). This suggests that a synchrony between the storyteller and the listener involves more than just the language of a story—it is a whole-person brain/body dance. Other research is beginning to show that people's brains synchronize strongly when they share the similar emotions suggesting that this may promote social interaction (Nummenmaa et al 2012).

It is important to note that the brain synchrony seen in this research is between strangers who are engaged in following a narrative or sharing an emotional experience. It is a short jump from a storyteller–listener communication to a teacher–student one. It seems reasonable to suggest that when students and teachers are engaged in a lesson, their brains are firing in synchrony. When we share a positive emotion in the classroom—a shared laugh, the excitement of discovering something new, or a personal story—everyone in the class has met the first of Fredrickson's trio "sharing of one or more positive emotions between you and another."

As if electrically synchronized brain/bodies were not enough, there is a biochemical dance that is occurring between connected people who experience positivity resonance, the second part of

Fredrickson's trio. It involves oxytocin, which some call the "love hormone," an oversimplification that contains a kernel of truth. Oxytocin is present in large quantities in mothers when their children are born and for both partners during sexual intercourse. Recently improved measurement techniques make it possible to track the ebb and flow of oxytocin during normal day-to-day interactions. One typical daily event that raises oxytocin levels is telling an intimate story about your life. When you reveal an important aspect of your life to someone you trust, your oxytocin level rises and you tend to further trust the listener.

It is also interesting to note that while oxytocin is flowing, one is sparked to trust others, but this is not blind trust. Oxytocin sharpens the attention to the other person's eyes and body language, making it more likely you will see the clues of an untrustworthy person. We are astute in judging the other person's intentions when oxytocin is flowing, and we simultaneously feel more connected and less stressed. When you are teaching and use yourself as an example by sharing a story of your life—we all know the power of teaching stories—you are synchronizing your brain with your students' brains. This synchrony corresponds to Fredrickson's second aspect of her positivity resonance trio.

The third aspect of Fredrickson's love trio is "a reflected motive to invest in each other's well-being"; in other words, a commitment to caring. When asked, most educators willingly admit that they are invested in students' well-being, and many students respond in kind. This reflected sense of care for another's well-being is partially explained neurologically by Porges's polyvagal theory. It bears repeating within the context of understanding love that the ventral vagus nerve connects different parts of the brain stem to the internal organs, particularly the heart. It also affects the muscles around the eyes. When oxytocin is flowing, that is where we focus our attention. Even the tiny muscles in the middle ear are tuned to help you pick someone's voice out from background noise. The vagus nerve regulates your heart and breathing so you can measure the vagal state. When you inhale, your heart rate rises slightly so your blood can absorb as much of the fresh oxygen breathed into your lungs as possible. When you exhale, your heart rate drops slightly because your lungs now have less oxygen to pass on to your blood. This heart rate change is the measurement of the "vagal tone"; the bigger the rate of change, the better. A high vagal tone shows that your brain is flexibly regulating your heart rate. People with high vagal tone are also more flexible and adaptive in their response to a wide variety of events, have better interpersonal skills, and are more adept at making positive connections with others. Fredrickson highlights the most important point of vagal tone for educators: "Compared to people with lower vagal tone, those with higher vagal tone experience more love in their daily lives, more moments of positivity resonance" (Fredrickson 2012, p. 56). So vagal tone is a nonconscious process that affects how humans connect to other people, including students—and their vagal tone is related to how they connect with us, hence the neurology of Fredrickson's reflected motivation for caring for the well-being of others.

Summarizing

Teachers and administrators are swimming in an invisible ocean of relationships. Knowing about human attachment and love can make for more effective teachers and make the days in the classroom go more easily.

It is now known that in the first few days, months, and years after birth, children's brains are structured by their connection to other humans, especially their parents. Good enough parents comfort their children when they are anxious, so the kids learn to rely on adults to comfort them when they are afraid. They are more likely to react to teachers in a manner that shows they expect to trust them and they will be more likely to weather the emotional and relational storms of normal schooling. Children that have grown up with less than adequate parenting will respond to teachers either with a lot of anxiety, worrying they will not be there when they need them, or they will dismiss teachers as irrelevant and make a minimal effort at connection. There may also be a few who have had parents who have hurt them (or who have high levels of anxiety). These students will react to teachers in a wide variety of intense ways.

The love we have for students, regardless of their attachment style, is based on innumerable daily brief instances of positivity resonance with them, times when our brains and neurochemistry have been ebbing and flowing in a synchronized dance and our social engagement system, the ventral vagus nerve, has been activated. Who knew that the simple heartfelt statement, "I love these kids," would have so much neuroscience supporting it.

Tools for School

1. Savoring. This "refers to our awareness of pleasure and our deliberate attempt to make it last" (Peterson 2006, p. 69). Savoring can be used to purposely focus attention on moments of positivity resonance. They are fleeting and easy to dismiss as unimportant, but they are at the core of the connection and love teachers feel toward students. If one dismisses them as unimportant, one misses an opportunity to add some positivity to one's day. The exercise of savoring offers us a way to increase the positive effect of these brief moments. The first step is sharpening your perception. Reading about these moments can sensitize you to being alert for their appearance during the day. Soon after you notice one, make a mental recording of the details so you can remember it clearly later (e.g., Who was the student? What did he or she say? What did I say and do? When did it happen?). Next, as you remember it later, let yourself focus on and absorb the moment by feeling the connection and the enjoyment of it. No multitasking allowed; instead, just sit with the feeling of positive resonance. Then congratulate yourself for being a good teacher and forming positive connections to your students. Finally, share as many of the details as possible with trusted colleagues. Taking a few minutes to savor these and other positive moments at school can be used to amplify their positive effect on you. Sharing them can help colleagues remember and learn to savor their own.

2. Disconfirming Experiences. As the words suggest, disconfirming experiences surprise insecurely attached students because they run counter to the way they have learned to expect relationships with adults to unfold. If you suspect that a significant number of your students are anxiously attached—remembering that the research suggests it will be between 20 and 50 percent (Bergin and Bergin 2009)—then consistently responding positively and firmly creates multiple disconfirming experiences because these students have had early lives of inconsistency. Some of the ideas will sound familiar because they are sound teaching techniques. For example, beginning every class by describing what you will teach, teaching it, and ending each class with describing what you taught is a familiar technique. Repeated daily, it helps create a predictable structure for students who have had unpredictable lives at home.

Another disconfirming experience, the mindful moment, is described in more detail in Chapter 7. As a beginning exercise, starting a class with one or two minutes of quietly focusing on the breath, simply counting from one to six on the in-breath and one to seven on the out-breath may make a remarkable difference in students' ability to be present. This simple technique lowers stress, opens students to new learning, and calms students and teachers after the busy-ness of the hallway, cafeteria, or teachers' room.

A fun disconfirming technique that helps students with anxious/ambivalent attachment style is to demand that they make at least one mistake during the class. We all know that new learning involves making mistakes, but anxious students often emotionally overreact to errors. Asking for mistakes will have a paradoxical effect on them (expect confused looks and nervous laughter at first). Some teachers also ask students to correct them when they make mistakes (and then purposely make them). This also improves student attention as they listen carefully and try to be the first person to catch the mistake. It also creates a safe environment for the anxious student as teachers role model an open and accepting response to mistakes.

3. Discover Your Own Attachment Style. There is some controversy about assessing the attachment style of adults because it is more complex than assessing a toddler's. With many different attachment experiences encoded in the brain, an adult's style may vary depending on the context. For example, an adult may have one style with a romantic partner and a different style with a living parent, friend, or co-worker. That said, one of the most valid means of discovering one's attachment style is the Adult Attachment Interview (AAI) (Main, Hesse, and Goldwyn 2008). The AAI is a clinical semi-structured interview of about twenty questions asked by a highly trained interviewer over at least an hour, and it has extensive research validation to support its conclusions (Hesse 2008). As you gain greater awareness of your attachment patterns, you will more likely be able to understand the behaviors of students that often arise from their earliest experiences. This new sense of yourself can also help you spot those moments when your style is getting entangled with that of students, colleagues, and administration. Awareness alone can offer some flexibility of response.

4. Circles of Trust. This is a process developed by Parker Palmer, author of *The Courage to Teach: Exploring the Inner Landscape of a Teacher's Life* (2007). He suggests organizing regular meetings among faculty. These meetings have a specific structure designed to increase trust and connection. Membership in the group is open and voluntary, and there is a specific meeting length and end date. The group is led by a trained teacher-leader who makes it possible for group members to openly share their inner lives related to teaching. The groups as described by Palmer (2004), as well as my limited personal experience with participating in them, make it seem likely they would support the development of trust and secure attachment among a school staff.

Chapter 4

The Attentional Circuits of the Brain

THE CAPACITY TO PAY ATTENTION IS FOUNDATIONAL FOR THE ABILITY to learn. You can only encode new information into memory when you are consciously attending to the information at hand. Beyond learning, the capacity to shift attention from one focus to another has a major influence on the quality of one's life experience. If I have a headache, my mind may be drawn toward the pain, noticing its intensity and how it affects my overall feeling about the day. With well-developed attentional circuits, I might be able to shift my attention to a more pleasant focus, with the result that I perceive the pain in my head as less intense, and perhaps the quality of the day will improve. When teachers help students cultivate attentional capacity, they are teaching a life skill as well as laying the foundation for purposeful learning.

It is also helpful to acknowledge the complexity of this task. Although humans have some core circuitry in the brain that predisposes us to explore novel terrain (both physical and intellectual), that circuitry is also influenced by past experiences of exploration, by the quality of safety and connection we feel in the moment, by the match between the information offered and what matters to us, by the overall environmental circumstances, and by our general status that day—well rested, getting sick, needing physical activity, worried about a friend, falling in love? The list is unique to each person and endless. As a result, except on rare occasions, the quality of attention in the classroom is variable, like a flowing river. What I investigate here is ways teachers might bias the situation in favor of greater attention for most students most of the time.

I begin by inviting you to pay attention to some information about the circuitry that allows you to direct your mind toward a particular focus and then explore some practices that can increase this capacity.

The Flashlight of the Mind

It can be helpful to imagine attention as a flashlight in the dark. When you walk with a flashlight at night, you have a choice about what aspects of the environment to illuminate and which to ignore. You use the beam to illuminate things you feel are a priority (rocks in the path, something you lost in the dark). By eliminating many possible targets of attention, you can focus with clarity and intensity on what matters most. This kind of self-chosen focus is called *top-down attentional control*. If a sudden movement in the dark startles you, without conscious thought, you reflexively and abruptly turn the flashlight to illuminate the potential threat. This is an adaptive response to novelty and is a second form of attending called *bottom-up attentional control*. As you will see, both forms of attention are valuable in the classroom.

Top-Down Attentional Control

Top-down attending, sometimes called executive attention or effortful attention, is consciously controlling the flow of attention based on your own priorities or those of others. You are using it now as you read this sentence. Research suggests that the brain's top-down, conscious attentional control involves a network of neurons in the prefrontal cortex and parietal lobe. This system functions as a priority map that guides the direction of an individual's flashlight of attention. "A priority map is a representation of the visual world in which items, objects or locations are represented by [neural] activity that is proportional to their attentional priority" (Bisley 2011, p. 54). Bisley goes on to state that the priority map is a combination of bottom-up input (new sensory information) and top-down influences, such as the goals of a task, evaluation of its importance, and personal bias. In other words, you do not randomly scan the environment, the priority map causes you to search for some features and ignore others. For example, if I am looking for a blue pen on my messy desk, my neural priority map will highlight features of objects like the cylindrical shape and the color blue by putting the specific neurons that see the cylindrical shape and color blue on high alert. I can more efficiently find my pen among the mess. In the process, the priority map will likely suppress my response to, for example, cup-like objects, so I might knock over my now cold cup of coffee as I search for the pen. I can also notice that I have a goal of finding the pen, and to a greater or lesser degree believe finding it is important. The length and intensity of my attention are influenced by the strength of these factors.

The mind can alter the brain's priority map in at least three ways: you can change your own map by choosing to focus on something different; you can be distracted by bottom-up influences so that your attention is shifted for you; or your map can be altered through your relationship with another person when he or she suggests directly or indirectly, that you focus on something in particular. Let's focus on this last kind of change, because it is the most central in the classroom. (You might take a moment to notice if you were able to shift your attention and let go of the first and second kinds of change when I suggested you focus on the third one.)

As discussed in the last chapter, if you have a quality connection with your students, the doors to learning begin to open. Connection brings increased safety so that more of the students' attentional capacity is available for being directed toward the priority maps the teacher suggests. Then, whatever techniques you use, you have the advantage that your students are more likely to choose to attend to whatever you offer because of the importance they place on their connection with you. The positive connection increases their attention on what you present in class. In reality, through the school day, this is the dominant dynamic: you offer curriculum and request that your students choose to attend to it rather than the thousands of other possible targets of their focus. This chapter looks at some effective ways to make that request, knowing that being within the context of a web of positive relationships helps students.

Bottom-up Attentional Control

Bottom-up attention is unconsciously triggered by stimuli such as rapidly moving objects in the periphery of your vision or sudden noises that might signal a threat. Equally, a person of interest strolling by or simply the beauty of the spring day outside the window can shift student attention from the bottom up and repopulate their priority maps. So can a sudden upsurge of worry inside their minds. Their attention is drawn toward the internal or external stimulus without conscious permission, before the prefrontal cortex is able to activate. The signal originates with the sensory or the memory circuits, which are primarily situated near the bottom of the brain, then relay the information up to the cortex (thus, *bottom-up* processing) (Buschman and Miller 2007; Grabenhorst and Rolls, 2008; Bisley 2011). For students and teachers, there will be a constant pull toward these nonconscious shifts as inner and outer experience unfold around and within you. How, then, might you harness this powerful form of attending and link it to top-down attending to enhance students' learning?

Response to Novelty

One kind of experience that grabs the bottom-up attentional circuits of the brain is novelty, and this can be both a significant challenge and powerful ally in class. Let's look at the challenge first.

The brain continually and adaptively assesses the unfamiliar for safety or danger, and quickly shifts attention toward what is new and unexpected (Siegel 2012a). Students and teachers will have varying sensitivities to this assessment process, which is initiated by, among other things, temperament and previous history. Some people are born with fear-monitoring regions of the brain that are more sensitive than they are in others. Based on research, temperament expert Jerome Kagan (1994) says that shy children may have highly active amygdalas as part of their genetic temperament inheritance. This means that new experiences more easily move them toward sympathetic activation and often promote withdrawal from novelty. However, when adults are able to understand the nature of this type of child's struggle, and gently encourage him or her to explore the world a bit, these children become indistinguishable from their less shy compatriots, even though their sensitive amygdalas continue to fire strongly. Once again, you can see how the power of interpersonal resonance with students enables them to access more of their potential and why a felt sense of safety is so important in the classroom.

Another source of sensitivity to novelty comes from early history—your own and that of your students. If one has lived in frightening conditions, the limbic circuitry develops patterns of expecting danger and, in fact, filters each new experience through this anticipation. Some seemingly distracted students may be experiencing bottom-up reactivity to stimuli that seem neutral to the teacher but are perceived as threatening to students because of their past or sometimes current circumstances. When you can learn to see the potential invisible causes for things that may look like behavior problems, you become more able to meet your students with empathy and wisdom. This makes room for the connections that can in time ameliorate some of these sensitivities. You may even recognize similar sensitivities in yourself and can use that self-awareness to inform your approach to your students. The process may gradually rewire the brain into patterns of safety and security.

Although bottom-up attentional processing can cause challenges in the classroom from students' attention being grabbed by movement seen through the window to a traumatized students' fight-or-flight response being triggered by a sudden noise, bottom-up processing can be useful. Many teachers unknowingly adopt techniques that use bottom-up processing to grab attention. For example, I have observed teachers use sudden voice changes, humorous turns of phrase, sudden animated movements, silly hand gestures, goofy hats, odd clothing, and sometimes well-planned lessons that include a bang. A chemistry teacher I know used dramatic experiments to capture student attention. (There is nothing like blowing something up to activate teenagers' bottom-up attentional systems.) The teacher realized that the bottom-up nonconscious attentional focus, once stimulated, can lead to the conscious top-down attentional focus needed to complete a lab report. Adopting techniques like these can help you use bottom-up processing effectively in the classroom. You can develop techniques that work by keeping in mind that novelty activates these attentional circuits.

The SEEKING System and Heightened Attention

We can harness an ancient neurological system deep inside students' brains—the SEEKING system (researcher Jaak Panksepp capitalizes the term to distinguish it from ordinary usage of the word). You may recall that as the human brain evolved, older structures were maintained and newer structures added on top. As a rule of thumb, you can judge the evolutionary age of a brain structure by its location; the lower it is in the brain hierarchy, the older it is in evolutionary terms. Novelty seeking may be an ancient addition to our brains because it involves structures in the brain stem, a region located near the lowest point of the skull, just above the top of the spinal cord.

The SEEKING system is one of seven emotional-motivational circuits located deep in the brain stem. It pulls attention outward, toward new adventures (Panksepp and Biven 2012). You have experienced the SEEKING system yourself: think of the anticipation you felt before a wonderful meal and maybe a glass of wine on a Friday night after a hard work week; the sweet anticipation of awaiting the return of a loved one after a long separation; the anticipatory eagerness before the beginning of summer vacation. These are all examples of the SEEKING system at work (Panksepp and Biven, 2012).

Louis Cozolino, in his wonderfully well-researched book on education, says it well: "Human animals are generally considered to be exploratory creatures and we are rewarded for our curiosity by the generation of dopamine and endogenous opioids, which are stimulated in the face of something new" (2013, p. 164). In other words, first you get the good feeling that seeking brings as you move toward something new and then the equally good feeling of finding when you get what you have been seeking. These circuits lead some to think outside of the box, enjoy solving new problems, and pursue adventures of all types. The attention-grabbing effect of novelty is so built into humans that it is evident in babies as young as four months, which is when they begin to prefer unexpected sounds to familiar ones (Cozolino 2013). We might speculate on the evolutionary functions of the SEEKING system; its bottom-up attending to novelty could have been a means to motivate early humans to investigate new territories for food and shelter. The narrowing of attention in response to unexpected changes in a familiar environment could have been a useful defense in signaling the presence of a predator or other danger.

Schools often unknowingly stimulate the SEEKING system during the run-up before large sports events, for example, a football game against a long-term school rival. You can use the system to your advantage by purposely creating unexpected changes in the classroom or raising students' anticipation of an event like the chemistry teacher did days before his legendary explosive experiments.

Looking Deeply into the Bottom of the Brain

Let's look more closely at how the SEEKING system collaborates with other affective circuits. Animal research has revealed six other affective circuits in addition to the SEEKING system. Humans share these systems with all mammals.They operate nonconsciously and originate near the very bottom of the brain.

- CARE/BONDING system: "the impulse to envelop loved ones with gentle caresses and tender ministrations" (Panksepp and Biven, 2012, p. 36). Without this system, taking care of the young (including teaching them) would be tedious drudgery; with it humans can feel a state of deep connection and satisfaction when caring for children. This system is also related to the experience of love.
- PANIC/GRIEF system: when active, this system creates an internal experience of intense psychological pain. Social connection alleviates this pain and replaces it with a sense of comfort and belonging. Feelings of love also include fear of losing the loved one, hence the PANIC/GRIEF system is part of love.
- FEAR system: this part of neurology creates the negative affective state that all mammals want to avoid. The details of the effect of the FEAR system on the brain and body were discussed previously.
- RAGE system: "Causes animals to propel their bodies toward the offending objects, and they bite, scratch, and pound with their extremities. Rage is fundamentally a negative affect" (Panksepp and Biven, 2012, p. 36).
- LUST system: when this system is active in mammals "they exhibit abundant 'courting' activities" (Panksepp and Biven, 2012, p. 36). Likely with the simultaneous activation of the SEEKING system, it can lead to the pursuit of a mate. The LUST system is obviously related to love.
- PLAY system: "is expressed in bouncy and bounding lightness of movement, where participants often—poke and rib—each other" (Panksepp and Biven, 2012, p. 37). In animal experiments, there is an alteration of dominance and submission. As long as these two roles change, the play continues; when one continually dominates, the play changes and becomes

negative. The PLAY system is one of the main sources of friendship.

These systems work together, for example, when you feel interpersonally connected, the circuitry of PLAY and CARE/BONDING come online along with SEEKING (and LUST, which arrives in adolescence). Playfully seeking new knowledge uses these coordinated systems and is a beautiful thing to see blossoming in our classrooms.

FEAR, RAGE, and PANIC/GRIEF show up along with SEEKING during times of disconnection. One of the signs of PANIC/GRIEF, arising from separation distress, is clinging—to the teacher or to peers. Educators often don't tolerate clinging very well, but perhaps when you understand that it can be caused by an ancient neural system present in all mammals, you may be able to meet the child's need long enough for him or her to regain a feeling of connection, which will ultimately stop the clinging. When you experience disconnection, your SEEKING resources get narrowly directed toward regaining connection, and once you are connected, you are free to explore widely and creatively.

As you can see, all roads lead back to the central importance of secure connections in the classroom as a prerequisite for new learning.

The seven emotional-motivational systems described by Panksepp and Biven appear to be hard-wired in the human brain at birth. They are nonconscious "instinctual emotional responses that generate raw affective feelings that Mother Nature built into our brains. We call them '*primary-process* psychological experiences'" (Panksepp and Biven, 2012, p. 9). SEEKING directs a "persistent exploratory inquisitiveness—approach and engagement in the world" (Panksepp and Biven, 2012, p. 36)—which is exactly what you want for your students. This is the circuit that underlies human curiosity, but curiosity also involves the use of higher brain regions, for example, the limbic region and the cortex (Panksepp and Biven, 2012). Curiosity may be thought of as a higher-level emotion that is built on the foundation of the ancient SEEKING circuit.

All of this suggests that the more your lessons are novel in their content or presentation, the more likely you will be to give the already active SEEKING system a focus for its powerful energies, and in the process, harness the effortless resource of bottom-up attention. Students will not need to engage the more effortful top-down attention as their primary way of engaging with the material.

Choosing to Use the SEEKING System

A colleague once shared with me a story about teaching first-year junior college students their dreaded English 101 class. As a new teacher, she followed the previous teacher's syllabus (the first difficulty, she later realized, was that she didn't have any excitement about the material in this format) and found herself struggling to keep her students' attention on the task of learning to write well. By her third year, she was so discouraged that she faced a crisis of whether to continue in teaching, but she decided to draw on her background in drama to do something different—novel for her and novel for them. She assigned a short play by Eugene Ionesco, *The Bald Soprano*, and divided the class into three groups. She assigned the groups to prepare and put on a performance of the play without having any contact with the other groups. The writing assignments were all attached to their experiences—from research about Ionesco, to reflections on their experience of performing, to commentary about the various performances (including philosophical exploration of whether the play existed on the page at all or only in the performance). After the expected initial wave of bottom-up fear about performing, the groups settled into working out which member could do which task most easily (choosing tasks based on their strengths) and then collaborated on a number of the written assignments.

The class almost effortlessly shaped itself around this novel approach without a great deal of top-down structure. Although the cortex was clearly recruited to do the necessary scholarly tasks, it seemed as if the novel approach also activated the students' SEEKING system. My friend continues to remember the excitement this novel approach engendered in her students and herself, as well as its marked contrast to the two previous years of teaching from a mundane and predictable syllabus. It might be possible to let your mind run wild in this way, too. See what you can dream up that excites you and builds on your students' and your own inborn response to novelty. Introducing novelty into a lesson gives the SEEKING system a target, engages the PLAY system, and makes room for some CARE and BONDING to emerge at the same time—which brings us back to the role of relationships in shaping attention.

Cultivating Attention in the Garden of Relationships

The emotional-motivational attentional systems are available and at their best when people feel connected to another, which means that learning is essentially a process rooted in relationships. Daniel Siegel put it well: "In parenting, teaching, and psychotherapy, we use the relational aspect of the mind to focus the attention of our child, student or client/patient to activate neural firing in a way that can potentially alter the structure of the brain. This is the fundamental basis of learning" (Siegel 2012b, p. 9-3). Remind yourself regularly that you are able to use your relationship with students to help their learning in two important ways. First, understand that anxiety and sympathetic activation make learning, curiosity, and exploration very difficult because of the natural and adaptive constriction of attention when one feels afraid. Anxiety can be ameliorated by a secure attachment to an adult. When you develop a supportive school and classroom culture that places a priority on relationships, you begin to lay the foundation for lowering anxiety and opening the neural pathways of learning. Second, within this context, you can positively influence your students' attentional priority maps to focus on important aspects of the lesson, in part because attachment circuitry prompts people connected with those who support them. Students are more likely to gravitate toward doing what the teacher asks because it maintains and strengthens their bond. Because attachment systems also thrive on harmonious connection, the interchange benefits brains and bodies as well. In Tools for School at the end of this chapter you will find some concrete ways you might put this awareness into practice.

I conclude this review of the benefits human connection offers to students' ability to pay attention by looking at something called social baseline theory (Beckes and Coan 2011). Beckes and Coan make the point that unlike many animals, humans are less adapted to our environment and must rely on other humans. "In our view, the human brain is designed to *assume* that it is embedded within a relatively predictable social network characterized by familiarity, joint attention, shared goals, and interdependence" (2011, pp. 976–77). These words seem to describe an ideal teacher–student relationship, so it is quite heartening to imagine that the capacity for it is built in because humans may have a genetic preference for such collaborative environments.

How does this influence students' capacity to learn? One of the discoveries in the research in question is that when we are in proximity with people we trust and with whom we've formed a bond, we require fewer resources from the prefrontal cortex to remain emotionally regulated. Rather than having to expend energy to become calm, we instead appear to return to a baseline of calmness that arises when we are connected (Coan, Schaefer, and Davidson 2006; Eisenberger et al. 2007; Coan 2010). This means that humans have significantly more neural real estate available to focus on learning. Also, when facing challenging tasks, if you have someone you trust at your side you perceive the task to be less difficult—the more positive the bond, the greater the change in perception (Schnall et al. 2008).

You can imagine that students sensing their teachers are their nonjudgmental, supportive allies who are also invested in them learning could have a profound effect on their openness to new information and willingness to take on more difficult challenges—both of which increase the likelihood of them attending well. All of this points to how you can collaborate with the natural processes of the brain to create environments in which both teaching and learning are pleasurable and effective.

Effortless Attention

Let's look briefly at a third kind of attention—a state of flow. Research begun by Mihaly Csikszentmihalyi (1990) and expanded on by others has focused on an experience you may have had, one that would be nice to create for students: "when people enjoy most what they are doing—from playing music to playing chess, from reading good books to having a good conversation, from working their best to trying to beat their own record in sport—they report a state of effortless concentration so deep that they lose their sense of time, of themselves, of their problems. We have called this the *flow experience*" (Csikszentmihalyi and Nakamura 2010, p.180-181).

Csikszentmihalyi has used what is called *experience sampling* (self-report many times a day about the current quality of the present experience) with several hundred middle to high school students around the country. He found a large difference in students' quality of experience between effortful and effortless attention. In a state of effortless attention, they reported being simultaneously more excited yet more relaxed than during times of effortful attention. They described a favorable match between the challenge of the task and their skills. When the level of difficulty of a task was high and individual skill level for the task was low, they experienced anxiety; when the difficulty of the task was low and their skill level was high, they experienced boredom. Another key to the experience was feeling the activities were freely chosen and generally were not productive activities (e.g., not schoolwork).

Csikszentmihalyi also discovered that flow is more likely to occur when people are using well-learned routine skills that have become automatic because less attentional effort is required, so people in the flow state use the conserved attentional energy to concentrate more on the task. Imagine a guitarist who has mastered her instrument well enough so she has to pay little or no attention to the mechanics of playing and can thus use her attentional energy to focus on the music itself, and hence enter a state of flow. Using the same line of logic, an elementary schoolteacher might be able to imagine a student who has mastered basic single-digit addition entering a state of flow as he uses this skill to do basic accounting for a school store.

At the other end of the spectrum, Csikszentmihalyi's research with teenagers confirmed his results with adults: "on the whole when people say they are concentrating their attention they also report that doing so is hard, [and] their quality of experience is generally unpleasant" (Csikszentmihalyi quoted in Bruya 2010, p. 184). This may be because high levels of attention are usually reported when people are doing something they are required to do by some external force, such as concentrating in school.

However, all is not lost for educators. Csikszentmihalyi then studied the students who experienced effortless attention at school and found a few common precursors that offer useful tips for teaching. When students experienced a state of flow at school they felt that they were in control, felt more involved in the schoolwork, were not self-conscious, and were interested in the content of the material. Research on the state of flow completed over approximately thirty years has shown that three conditions need to be present for someone to experience a state of flow: clear goals, immediate feedback, and as already noted, a balance between the demands of the task and the skills of the individual. Teachers can sometimes reproduce these conditions in school. An example may help. In my experience, there is a considerable amount of enthusiasm by some students and staff for rock climbing and other challenging outdoor experiences. This appears to be true in spite of this valid question: why would anyone purposely climb a cliff and put themselves in danger for no particular reason? It is because rock climbing has the potential to offer all three of the conditions necessary to experience the state of flow. There is a clear goal, which is not the relatively distant goal of reaching the top but the short-term goal of finding the next handhold or foothold on the rock face. Then there is the immediate feedback of seeing that each handhold or foothold brings the climber closer to the top. The third condition, the match between the climbers' skill and the difficulty of the rock face, depends on the instructor assigning the correct level of challenge to the students. If it is too hard, the climb will be frustrating and dangerous, if it is too easy, it will not engage the attention of the climbers.

You can keep these three conditions in mind as you design classroom lessons. For example, create lessons that carefully consider how the challenge of the task meets the skill level of your students.

Offer short-term goals with immediate feedback and some way that students can check the accuracy of their answers immediately after completing a problem. I have observed a math teacher teach Algebra I using manipulatives, where the number and position of the blocks gave students immediate feedback on the accuracy of their work. The challenge of the lesson and the skills of the students matched well, because there was a minimal amount of frustration with the lesson. The students all appeared to be in a state of flow and they seemed surprised when the class period ended.

There is a fourth important factor for teenage students: offering choices of assignments so they can feel a sense of control over their work. The teachers I have seen do this well use methods such as project-based learning (Blumenfeld et al. 1991); or offer students multiple ways to demonstrate their learning of the information. For example, they might offer students a choice between writing a report, taking a test, or making a presentation to the class.

The four conditions—matching the challenge of the lesson to students' skill level, including short term goals, immediate feedback, and giving students' choices when possible—will help create a classroom culture that is fertile ground for the emergence of a state of flow and effortless attention. Keep in mind that even a few moments of this experience can create a zest for learning that can carry students through the more arduous parts of the work that require effortful attention.

What about ADHD?

With the best of intentions, for hundreds of years education and psychology have focused on what has been seen as weakness or pathology of attentional capacity. Rather than supporting strengths (the topic of Chapter 6), the unspoken assumption is that improving students' weaknesses or curing their ills will help them succeed. It comes as no surprise, therefore, that so much emphasis has been paid to the apparent pathologies of attentional control, such as attention deficit hyperactivity disorder (ADHD). The pathological approach has led many teachers to assume that all difficulty with attentional control is some kind of biological problem that can only be fixed with medication; for students with moderate or severe ADHD this may be partly true. If instead the attentional struggle is rooted in anxiety, concerns about what is going on at home, interpersonal difficulties at school, or the causes discussed earlier, the normal nonconscious response of the bottom-up attentional control circuits or normal sensitivities to some forms of novelty caused by students' early life experiences. Knowing these possibilities may be of help in finding ways to guide students back to attending. Researchers have known for some time that attentional control is at least partly a teachable skill (Meichenbaum and Goodman 1971; Buschman and Miller 2007).

Tools for School

To conclude this chapter, here are some exercises to develop attentional control and some teaching strategies that may help capture students' attention.

1. Teaching Students to Aim Their Flashlight of Attention. One exercise used by many outdoor educators as they venture into the field with their students is to simply pause and direct students' attention to their sense of hearing. Teachers ask them first to just listen for a few minutes, then list all the different sounds they hear. It is a simple five-minute exercise with the content area goal of identifying the sounds of nature students might not ordinarily hear. The potential side effect is a slight strengthening of the attentional control muscle. Each time you consciously direct your attention to something you choose (hearing, in this case), it is a bit of practice that helps strengthen the attentional control circuitry, making it easier to use the top-down attentional circuits to direct your attention where you choose. As we discussed previously, neuroplasticity occurs with repetition—neurons that fire together wire together. Giving students frequent practice directing the flashlight of attention to one of the senses (or to any target) will improve their skill at consciously controlling their attention.

2. Mindfulness. Mindfulness is another method of strengthening top-down attention. Chapter 7 has much more to say on this topic, but for a simple beginning exercise you might ask students to focus internally on their breathing. When they notice their thoughts wandering, ask them to gently bring their attention back to their breath. There is a good deal of research regarding mindfulness practices and attention (Siegel 2007). In my experience, spending just two minutes at the beginning of each class helps set a good tone, and the act of consciously bringing wandering attention back to the breath strengthens students' attentional circuitry. As you practice with them, you will likely find your own mind calming and your openness to your students expanding.

3. Influencing Students' Attentional Priority Maps. Earlier you learned that an individual's attentional priority map can be influenced by someone else. Some of us do this already. When you review/repeat the important information prior to a test, especially a midterm or final exam, you alter students' attentional priority maps to focus on and learn what will be asked on the test. A daily method used by some experienced teachers is to begin class by writing a few questions on the board, reading them aloud, and challenging students to interrupt the lesson when they detect the answers. The questions serve to influence students' attentional priority maps by directing them to search for the answers throughout the class time. This meets the criteria for creating priority maps: a combination of bottom-up input (new sensory information) and top-down influences such as the goals of a task, evaluation of its importance, and personal bias.

4. WWW. Explicit instructions to pay attention to specific aspects of class can change students' experience and has the potential to improve the classroom culture. You might end each class by asking: "What went well (WWW) in class today?" When asking the question, it helps to emphasize that although the answer needs to be specific, it does not need to be anything particularly important and can in fact be something that seems insignificant. Implementing this exercise may feel awkward at first because you are asking students to do something novel. At the same time, novelty has a good chance of catching their attention. Modeling this by offering your own WWW is a good way to help get students warmed up. Once you have established WWW as a class-ending routine, you can remind the students that they will be asked for a WWW at the end of the class. This will begin to orient a portion of their attention to the positive aspects of the day, creating a priority map that will allow them to have an answer at the end of class.

5. Pearls. Pearls is the nickname of another activity, one that is used in coaching and psychotherapy. Similar to the WWW exercise, at the end of the class or session the coach, teacher, or therapist asks each person: "What is one pearl of knowledge you have gained today?" In addition to pushing students to use their memory to review and summarize important information from the class, routinely telling them at the beginning of class that you will ask them for their pearl will set up an attentional priority map to seek out and remember bits of knowledge throughout the class.

6. Sharing Personal Information. Finally, adding more novel personal information to classroom lectures may be an excellent strategy to harness bottom-up attending and promote retention. Kintsch and Bates (1977) measured students' memory of a normal college lecture. They found that students had the best memory for extraneous information such as jokes and other personal comments by the professor. The best remembered information was unique in both content and emotion in comparison to the balance of the lecture, and one might speculate that the personal stories also gave the students more of a sense of connection to the professor. Imagine that a personal story regarding the content of a lecture with a classroom full of students with a secure attachment to the teacher would help cement the content in memory. For example, I still remember the personal stories shared by my eighth-grade social studies teacher of his work on the Panama Canal, and those tales were told more than fifty years ago!

Chapter 5

Explicit and Implicit Memory

EDUCATORS ARE CONTINUALLY INTERACTING WITH THE MEMORY CIRCUITS of students' brains, and recent research on the complex ways humans create and retrieve memories is improving our understanding of the process. Broadening our understanding of the neurobiology of memory provides the base knowledge necessary for creating lesson plans and interventions most likely to improve students' long-term retention. The neurobiology of memory also provides valuable insight into the behavior of students and staff alike. To begin the discussion, let's distinguish between two different forms of memory.

Two Functionally Different Forms of Memory

Memory research shows that there are two quite different forms of memory: *explicit memory* and *implicit memory* (Siegel 2012a). When most of us think about memory, we are usually referring to explicit memory—remembering in the present a piece of factual information or an autobiographical episode learned or experienced in the past. When you bring an explicit memory to mind, you consciously recognize you are engaged in the act of remembering. To encode an explicit memory, you must pay attention. Implicit memories are different from explicit memories in some significant ways: they don't require us to be paying attention to encode them; and we do not experience the memories as being in the past when we recall them. Instead of realizing we are remembering, we actively experience implicit memories from the past in the present.

When you experience implicit memories they seem body-centered, containing elements you could describe as emotional surges, body sensations, behavioral impulses, and perceptions. In addition, they often have sensory fragments embedded in them. For example, if the light was at a certain angle or someone was wearing a particular perfume, these small pieces of sensory information are encoded as part of the neural network of the implicit memory. They are more about the subjective aspects of an experience than the factual information surrounding it. Implicit memories can play an important unseen role in people's lives by nonconsciously influencing actions, feelings, thoughts, and relationships. For example, if your first experiences of school were warm and safe, then the next time you enter any school building, your body will experience these same feelings, creating a perception that this is a good place to be even before you have any experiences there. This can create a feeling of confidence, a body sensation of ease, and a behavioral impulse to move forward into this new adventure. This, in turn, influences how others respond to you. It would be difficult to overestimate the influence of these invisible but always active memories.

Implicit memories also create *mental models*, which are summations of repeated experiences that become expectations for future experiences. These mental models act as filters that bias our current perceptions in the direction of the already established model.

In other words, the mental model creates an expectation of what will occur next and can screen out, allow in, or color information depending on that expectation. On the other hand, mental models also make it possible to more easily live in the world; because of them you make nonconscious educated guesses about what is coming next, so not every experience is brand new. At the most basic level, by walking up and down stairs, you develop an implicit memory, an embodied pattern of activity that enables you to ascend and descend without much careful thought. If you did not have a mental model of stairs, every time you walked up or down would be a brand-new experience, like it is for a toddler, with similar results.

For teachers, as a normal part of creating implicit memories, your prior experiences with students as a whole have left you with explicit and implicit memories of students in general, biasing your view of them as a group before they even enter the classroom. Once they are in the classroom, you begin to form explicit and implicit memories related to particular students, so that repeated negative experiences can create negative expectations about those students. These can cause you to unknowingly screen out positive and neutral acts, while being sensitive to remembering negative acts. Teachers won't only remember what the students did (explicit memory), they also have a felt sense in their bodies of the influence the students' actions had on their subjective well-being (implicit memory). In truth, these implicit memories are what most powerfully create the perception that these students are difficult. It's the wash of uncomfortable feeling that comes over us every time we see or think about them, that sinking feeling you may have experienced when a certain difficult student enters your classroom. Sadly, this can create a self-fulfilling cycle because it screens out the student's neutral or even positive behavior, making it very difficult for him or her to change as well as making it difficult for us to see any positive changes.

The same thing happens for the mental models your implicit memory develops when you have repeated positive experiences with other students. You screen out their less than optimum behavior.

Because implicit memory highlights the negative aspects of some and positive aspects of others, the students on the receiving end of these mental models can respond to the way you treat them as unfair. To reiterate, this is the normal functioning of implicit memory and not a sign of poor teaching. However, as you become more conscious of this inevitable pattern, you might be able to purposely look for positive acts by students burdened with negative perceptions and comment on them. It is a variation of the behavioral intervention sometimes called "catching a student being good," but with the goal of changing your own mental model of him or her.

Realizing there are two types of memory, explicit and implicit, at work at all times in the classroom is an important starting point, and it will also be helpful for broadening your understanding of the memory process, beginning with a recently discovered inconvenient truth.

Explicit Memory Is Not a Hard Drive

Explicit memories are not videos, movies, or computer hard drive tracks recorded in the past and retrieved unaltered in the present. There are no "read-only" documents in the human brain. Much to the consternation of the legal and educational world, memories are fluid and changing. For lawyers and judges, this finding has created a nightmare by undermining the long-held veracity of eyewitness testimony; for teachers it changes the long-held commonsense understanding of memory. One reason metaphors of video recordings, movies, and computer hard drives are inaccurate is they confuse the three memory processes: encoding, storing, and retrieving.

Encoding Memories

The beginning of the memory process, *encoding*, is the initial activation of the brain's circuits and the trace this leaves behind, called an *engram*. For example, if I am teaching using a whiteboard on a Friday afternoon, complex networks of neurons are firing in my students' brains encoding engrams of the facts of my lesson (neuroscientists call this semantic memory) and much more. As the students look at the whiteboard, the visual neural networks, mostly in the occipital lobe at the back of their brains, are actively seeing and creating meaning from the marks I make; as they hear my voice, the networks of neurons hearing the sounds and making meaning of my words are firing in the temporal lobes on the sides of their brains; and, invisible to me, as their minds are excitedly making plans for the weekend, yet other networks of neurons fire encoding those emotional memories. There is even more going on in those brains, including encoding the frustration they feel from trying to understand my lesson (emotional memory), the pressure of the chair on their bodies (somatic memories), not to mention the image of students sitting in front of them, the colors of the walls, and so on. It is easy to get a feel for how much competition there is for the precious attention that is necessary to encode the memories of the lessons I am teaching.

If our brains encoded, stored, and later retrieved all the information available to them, we would live in a continual state of overwhelming confusion. The brain determines what is valuable information and stores it, and what is not valuable and ignores it. Before I walk through how the brain determines what is important and what is not, let's look at the details of the memory process itself.

Neurologically, memory is the probability that similar neurons active at the time of an experience will fire again at another time.

There have been innumerable research studies of memory, and Dan Siegel summarizes them this way: "The essential feature of these studies is that the connection of neurons in an intricate network (i.e., the structure of the brain) allows learning to occur. . . It is thought that the firing of single or collective components of a neural network alters the probability of patterns of firing in the future" (Siegel 2012a, p. 47). The increased probability of firing is the neurological equivalent of what we call memory. Donald Hebb (1949) discovered this process, and Carla Shatz (1992) authored the now famous quote "cells that fire together wire together" (Siegel 2012a).

There are two ways to alert the neural networks of students' brains that the information you are teaching is important and needs to be remembered:

1. Repetition: Each time a pattern of neurons fires, it makes it more likely they will fire in the same pattern again. Repetition signals the neurons and their synapses that the repeated information is important, and so it is more likely to be remembered.
2. Emotion: Input that occurs at the same time the person is experiencing moderate to strong (but not overwhelming) emotional arousal will be tagged as important. If the structures in the limbic region of students' brains, especially the amygdala, respond to the teacher and the content of our lesson, the information is more likely to be stored in memory; if the limbic region is inactive, the information will be tagged as less important and is less likely to be remembered.

Practice Makes Perfect

The cliché that captures the neurological reality is "practice makes perfect." Repetition, which includes but is not limited to rote memorization, works well for some topics such as learning the multiplication tables, memorizing a script for a play, or learning a poem by heart. Repetition is also important for skill development, such as playing a musical instrument, driving a car, or dancing. Teachers have used a variety of methods to help students learn in this way: flash cards for multiplication tables, practicing the same arithmetical operation using different numbers, conjugating verbs in foreign languages, and, in some schools, reciting aloud in unison. These methods work, but repetition has obvious drawbacks, including the drudgery of repeating the same thing over and over until it is memorized. It has the additional disadvantage of potentially creating implicit memories of the misery of the experience, and thus can work against a willingness to do more repetition. Fortunately there is another method for shifting information from transient neurochemical changes to long-term memory, one that involves emotions.

Emotion Tags Information as Important

Emotionally arousing experiences have a strong effect on the encoding, storage, and retrieval of explicit memories (Siegel 2012a). Knowing this fact opens opportunities for teachers. As noted before, the brain deals with an immense amount of information and must determine which pieces are important. The emotional center of the brain, the limbic region, continually makes nonconscious decisions about what matters. This part of the brain is constantly on the alert to keep you safe, so if a lesson seems dangerous, unexpected, or relationally appealing—the brain says, "remember it!" If it is expected, familiar, or relationally disconnected—the brain says, "don't bother." This means that you can influence how the emotional centers of students' brains respond to the information you teach, increasing the likelihood that the information will stick.

Most teachers focus on the content of the course and try to make it as interesting as possible. Unintentionally, they present the lesson in a way that alerts the amygdala, for example, bringing to light unusual or unexpected facts. There are many techniques to present course content, but let's shift the focus from the content of the lessons to the teachers who deliver it. In my experience, which is supported by neurobiology research, a combination of course content and the connection to and personal presence of the teacher is what enhances learning.

Mirror Neurons

There is a recently discovered neurological system that explains how the expression of emotions can influence students: *mirror neurons*. As you teach, your conscious actions and the nonverbal expression of your intentions and emotions create echoes of your state of mind within the brains of your students. Students resonate with teachers in a seemingly magical dance of synchronicity. The process begins with a special class of cells called mirror neurons.

The initial discovery of this system serves well to explain them. A lab in Italy was studying the motor neurons of monkeys and inserted tiny electrodes into the motor cortex of a monkey's brain. As the monkey reached for food, the electrode inserted in the motor neuron that controlled the physical act of reaching for something showed activity, which was an expected result. Then the unexpected happened. An experimenter in the room with the monkey reached for the food, and the same motor neuron fired in the monkey's brain while he watched. Why would the neuron controlling the monkey's reaching fire when he was only watching the experimenter reach? At first, the experimenters believed there was an error in their equipment, but finally realized they had discovered a new neurological system. Since then, these results have been replicated with other monkeys and in humans (Iacoboni 2008; Siegel 2012a). The results of the research indicate that the intentional actions of one person, controlled by a neural network in their brain, can activate a similar neural network in an observer's brain.

Mirror neurons and their connection to other regions of the brain create a rich circuitry of resonance. They enable an observer to imitate another's purposeful behavior and simulate his or her internal state—including emotions, body sensations, and intentions, at a minimum (Iacoboni 2009; Siegel 2012b). This breakthrough research has stimulated some controversy in the neuroscience community, and mirror neuron research is ongoing. Mirror neurons have now been found in other areas of the brain. Based on the current state of research, some believe mirror neurons could be a neurological explanation for empathy—the capacity to resonate richly with the inner lives of others (Iacoboni 2008; Siegel 2012a).

If you have ever watched a thrilling stunt, such as a tightrope walker or bungee jumper, you have already felt the effect of mirror neurons. As the person teeters on the wire or leaps into the abyss, you can feel your body trying to maintain balance or the bottom drop out of your stomach as you resonate with the intentional actions of the other. While this is a dramatic example, mirror neuron resonance shows up in more subtle but profoundly important ways in the classroom.

Mirror Neurons at School

Mr. Arker is a history teacher with a lifetime passion for politics, debate, and teaching. He reads voraciously and loves to discuss current events. With school colleagues, he usually brings up a news item and happily takes a stance that is further to the right than most of them. During his National Guard deployment to Iraq and Kuwait, he happily debated with his fellow soldiers even though he found his beliefs to be to the left of theirs. It is the debate that he loves.

Alice and John are two of his students, and I remember their complaints at the outset of the school year. Based on her previous experiences, Alice said, "Why do we have to take history? It's stupid and boring." John chimed in, "History is in the past. It's so done. It has nothing to do with me." Several months after the start of the year, I dropped by Mr. Arker's history class unannounced and was amazed at the palpable excitement as Alice, John, and the other students debated what seemed to be obscure aspects of Civil War strategy. When the students made a point, Mr. Arker would ask them for the facts to support it. As the emotions became intense, he diverted any personal attacks by asking them for the facts. I found myself swept into the debate along with the kids as it seamlessly shifted from the U.S. Civil War to the Mideast civil wars. Mr. Arker's passion for debate had captured all of us.

On reflection, I could see how he communicated his passion through his sometimes provocative, sometimes ironic voice tone; his animated facial expressions; and his purposeful hand gestures while moving around the classroom. His whole body exuded passion for the debate and its topic. It was obvious to any outsider that Mr. Arker had activated his students' (and my own) mirror neurons and their associated circuits to join him in his emotionally rich relationship with the material. He has a reputation among the students for giving lengthy, difficult tests. When I asked Alice if she was dreading his test, I was surprised when she told me she was worried but oddly looking forward to it. "There is something about his class. I can remember the stuff we talk about. Anyway, if I get something wrong, I can debate with him about it later." She went on to explain that when Mr. Arker gives back the corrected exams, he sometimes uses it as another opportunity for debate. If students can convince him with facts that he should add points to the score, he will. Mr. Arker is an example of how passionate teachers can create enthusiasm in their students through the mirror neuron system. It can take students from dislike to enthusiasm about a subject, showing how teachers' inner state as reflected in outer actions can increase the probability that those neural networks will fire again at exam time.

When teaching assignments are considered and class schedules are developed, it is important to answer the question: what are the teacher's strengths? What are her or his passions? The next chapter delves more deeply into the topic of teacher, administrator, and student strengths, because like Mr. Arker's teaching, strengths offer an opportunity to collaborate with the mirror neuron system to improve student learning. For now, imagine the benefits for everyone of working among a school staff who have learned their individual strengths, recognize their colleagues' gifts, and teach the classes they love.

Short-Term Stress and Memory

School itself can be stressful, and unsettling events or relationships outside of school can carry their influence into the classroom. The brain structure in the limbic region that is especially important for stress and memory is the amygdala. You might recall that the amygdala is the central hub of the brain's alarm system and is connected to various brain regions. You learned in Chapter 2 that when the amygdala is triggered, it ramps up the sympathetic nervous system, releasing adrenalin, norepinephrine, and cortisol into the bloodstream. At low levels of activation, the amygdala's alarm improves memory encoding by tagging the information at the time of activation as important, but there is an upper limit to this effect. Epinephrine, norepinephrine, and cortisol change the balance between explicit and implicit memory encoding. Epinephrine and norepinephrine increase implicit memory encoding, and cortisol hinders explicit memory encoding (Siegel 2012b). If the sympathetic nervous system becomes highly active, the whole organism shifts to survival mode. Remembering the lesson becomes less important than attending to the perceived threat, because attending to the threat will ensure basic survival.

Remember that student sympathetic activation is caused by the their perception of events, not the teachers'. Experiences that trigger students' amygdalas and sympathetic arousal may not trigger the teacher's, because of the differences in previous life experiences. For example, a student who has implicit memories encoded by repeatedly being shamed for difficulty in math may experience strong sympathetic arousal at the mere thought of being in an algebra class. These implicit memories may manifest themselves as inattention, fidgeting, talking with other students, or gazing out the window to try to calm the system. Understanding that the whole process may be out of the student's conscious control because it stems from an implicit memory can help the teacher modulate her response. Knowing that student behavior may be caused by triggered nonconscious implicit memories may help you turn to techniques that help calm the amygdala and sympathetic nervous system, rather than simply doling out punishment. This is especially true if you can begin to sense that being angry or punitive with the student will only further activate his fight-flight-freeze system, making it harder for him to focus on learning. The goal is to lower the level of stress so that once calmer, students will be able to direct their flashlight of attention toward the lesson and engage their explicit memory to remember what is being taught.

Long-Term Stress and Memory

As you have seen, amygdala activation and the accompanying elevation of the sympathetic nervous system are designed to ready the body to meet the needs of short-term stress, with the system returning to a quieter state when the perceived threat is over. However, there are multiple detrimental effects if the alarm system stays active for an extended period of time, even if it is at a low to moderate level. Ongoing stress or trauma creates sustained sympathetic activation and prolonged high levels of cortisol, a chemical that, if chronically present, can damage the nerve tissue of the hippocampus, the hub of the brain's memory circuits (Cozolino 2013). The hippocampus is a limbic region brain structure, and there is one in each hemisphere. They extend back from the amygdala and are shaped like seahorses, from which they draw their name. They are involved in encoding and storing all new explicit memories and retrieving them until they go into long-term storage. They also help make connections between previous learning and new learning (Siegel 2012b; Cozolino 2013). When you design a lesson to build on previous learning, you depend on the hippocampus. Well-functioning hippocampi are crucial to all phases of taking in new information.

Being aware of the effects of single-incident traumas (school shootings, sexual assault) and ongoing stress (chaotic families and neighborhoods, poverty, other siblings in trouble, parents too busy to attend to their children) can help teachers develop a compassionate understanding of why some students struggle so much with the tasks of learning. In addition, you can help them generate new neurons in their hippocampus. Research suggests that new learning and physical exercise stimulate neurogenesis (Van Praag, Kempermann, and Gage 1999; Aimone, Wiles, and Gage 2006). Creating a safe, relationship-rich classroom culture can make it a positive island in an ocean of stress for students. As their nervous systems calm, new learning becomes possible and neurons connect and grow. When we send them out for recess or sports, we can also celebrate the rebuilding of the hippocampus. Gym teachers and coaches can be our partners in improving students' memory system (Ratey and Hagerman 2008).

Memory Retrieval

Once students' brains have encoded and stored explicit memories of the lessons, there comes the issue of retrieving what has been stored. Memory retrieval, like encoding and storage, is not as straightforward as you might think. An individual's ongoing feelings (e.g., long-term depression) and current affect (e.g., feeling a bit blue because it is a rainy day or frantic because of a faltering friendship) have an effect on how easily he or she will retrieve previously stored memories. If students' current mood matches their mood at the time they encoded the memory, it will help them recall the information, something called *state-dependent learning*. It refers not only to our mood when hearing the new information but also to the sensory experiences occurring at the time of the original learning. As you recall, what fires together wires together, so both mood and sensory input are encoded in the same complex neural network as the information in our lesson. When there is a match between any of those aspects and the current situation, we retrieve information more easily. After a professor friend explained this principle to one of her graduate students, the student put a dab of perfume under her nose when she attended her very difficult anatomy class, did the same on test day, and reported excellent results.

It may be obvious at this point that previous test-taking experiences can strongly influence the ease with which students recall needed information. As stress escalates, retrieval improves but then quickly declines if the stress increases. Because there is so much pressure on students to produce good test results, it can be challenging to remain in a calm, open state about exams. Chapter 7 on mindfulness may help both students and teachers with that particular difficulty.

Encoding, Storing, and Retrieving Implicit Memories

Just as with explicit memory, people encode, store, and retrieve implicit memory, although the process looks very different. An example may help clarify.

You may remember Gary from the brief description of his interview in Chapter 1. He had been expelled from his local public school because he got into a fight in a classroom, and during the fight hit a teacher, splitting his lip. I first met Gary about two weeks before the fight occurred, when his parents and the public school staff called separately to refer him to my private practice. There was a stark difference between the Gary his parents described at home and the Gary the staff knew at school. Gary at home was kind to his brother, voluntarily helped an elderly couple, and loved hunting and fishing. The school staff described a boy who was often aggressive and always angry, bristled at any teachers' requests, and often joined fights between other students, making them much worse. Not surprisingly, Gary's parents mistrusted the school's description of their son, and the school staff believed the parents minimized Gary's anger and violence.

I was asked to attend an emergency meeting at his school a few days after the classroom fight that injured the teacher, coincidentally arriving in the parking lot at the same time as Gary and his family. We chatted as we walked toward the school building, and Gary was reasonably calm and pleasant. As we entered the building, I noticed an abrupt change in his voice tone as he angrily muttered, "I'm pissed! I get pissed off when I walk through that door." His insight certainly was correct. "I don't get it. This f***ing place, I hate it." He was belligerent throughout the meeting, making it much worse for himself and stressful for all in attendance.

After we left the school building and were back in the parking lot, he apologized for his anger and seemed genuinely remorseful. "I can't help it. I get so pissed in there." Over the next weeks and months, the narrative of Gary's school life unfolded. He described the relentless bullying he suffered, accompanied by the emotions and body sensations from those times. He sobbed as he angrily described being a target of bullying through his middle school years. He told me about his impotent anger in the face of the mocking, as well as the fights he lost. His resentment and anger were present as he proudly described his growth spurt during the summer before entering high school and the confrontation with his primary nemesis in the hallway at the beginning of that school year. The confrontation stopped the bullying, but he could not stop himself when he saw other students being bullied. "When I see them making fun of one of the geeks, I can't stop myself. I get so angry I attack. Sometimes I don't talk—just walk up to them and smack! And I don't even like geeks!"

Humans encode implicit memories mainly in an embodied way, storing them as emotional memories with behavioral impulses, perceptions, and sensory fragments tied to them, and retrieve them based on internal and external experiences that activate them. All of this happens largely outside our conscious awareness. When a painful or frightening implicit memory is retrieved, it takes away conscious choice. These fast-acting memories take over emotions, thoughts, and actions so rapidly that you can become bewildered. In Gary's case, he had made strong implicit memories of the pain and fear of being bullied. The sensory experience of entering the school changed his perception from safety to danger and set in play a chain reaction of implicit responses that led from the emotions of pain and fear to the behavioral impulses of rage and violence. Without realizing it, Gary was retrieving the emotions and actions in the present that had been encoded and stored in the past when he was being victimized. These retrieved implicit memories led to his violent actions and his sense that they were uncontrollable.

Once we understood that his actions were based on an unleashed implicit memory, we realized that a new school placement might be the answer. As you might recall, implicit memories can be altered by disconfirming experiences, so the new school set out to create those for him beginning on his first day. As was the custom at the new school, staff authentically welcomed him along with all the students each day. When he arrived with a large iced coffee (the school allows only water), he was asked to dump out the unfinished coffee before entering the building. His anger was met with kind, consistent explanations and extra time to finish the drink outside the building. He then dumped the extra coffee on the flower garden outside the door. A passing comment by a staff member turned the daily coffee dumping into a multiyear botany experiment. We all carefully watched the effect of the coffee on the plants. (They actually grew larger than the others in the garden.)

In addition to these disconfirming experiences, the staff taught him how implicit memory works so that he would not be so baffled by his own mood changes and the memory of violent behavior in the previous school. Such explicit explanations can help students gain a clear semantic understanding of the nonconscious implicit memory retrieval process. This understanding helps make the confusing and sometimes overwhelming implicit experience comprehensible. "Name it to tame it" is a phrase that describes the process of adding semantic information to nonconscious or emotional experiences, a process that can ameliorate their effect (Siegel 2010a, 2010b).

Gary never had a violent incident at his new school and went on to become a student leader, often helping younger, more vulnerable students. It also won't come as a surprise that his grades improved once the implicit stress was understood and relieved. Implicit storms make it almost impossible to learn, and calming them by understanding the causes and providing supportive interventions and caring relationships often allows the SEEKING system to once again orient toward learning.

Thought Experiment

To have a small personal experience of your own implicit memory, read to the end of this paragraph and try this thought experiment. Begin by checking in with yourself. How do you feel now? Tense? Bleary-eyed from reading? Happy and relaxed? Now, take a moment to breathe to help turn down your sympathetic nervous system and turn up your ventral vagal system—slowly count from one to six as you inhale and slightly slower from one to seven as you exhale. Do this for at least three full inhale-exhale cycles. Once you feel calm, let your mind float back to a specific time when you were at your best while teaching. Spend a few moments with the details of this explicit memory. Who was there? What did you do? When did it happen? Where did it happen? How did it all unfold? Once you have a clear explicit memory with all the details fully in mind, savor the memory for several moments. Try this now, before you read further.

As you open your eyes, notice how your body feels now. Do you feel a sense of warmth and goodness in your chest or belly? A surge of warm feelings of connection? Even though you explicitly know that this experience is in the past, you can notice that the bodily sensations and emotions are with you right now, in the present moment. These sensations are probably part of your implicit memory of the event. Regardless of whether we have positive or painful implicit memories, when they are retrieved, they are experienced as though they are happening now.

As you become more acquainted with your own implicit processes, it can help you see them when they are arising in your students as well. This gives you the creative edge to craft disconfirming experiences rather than taking a stance and actions that only confirm students' past negative experiences.

Tools for School

1. A Memorable State of Mind. Here, you could succinctly say that anything you can do to create a learning environment that is rich in safe relationships will lay the foundation for better initial learning and storage and make it more likely that students will be in a state of mind to attend, encode memories, and retrieve the material as needed. Your clarity about the interplay of implicit and explicit memory can help you be curious about why certain students struggle to learn at a level commensurate with their apparent potential. Warm curiosity, coupled with a growing understanding of the brain's processes, opens a door that judgment and punishment closes. Although it may seem odd that a change in perspective might be the key to increased memory retention, the chapter that follows on a strengths-based approach to student learning covers that in more detail.

You know that implicit memories can cause difficulties for students who have had negative experiences at school, for example the big trauma of being bullied or the multiple smaller traumas of struggling hard only to fail. Implicit memory can also cause difficulties for students and teachers who have negative experiences at home, for example with abuse or parental alcoholism, poverty, or mental illness. Implicit memories can make students behave in a negative manner in the classroom. Before you decide they are doing something purposefully negative and should be punished, consider that they may be experiencing an implicit memory. One way to intervene to help students cope with the effect of implicit memories is to use the 1 to 10 scale described in the Tools for School section of Chapter 2.

2. Repetition. This tried-and-true tool is well known to all teachers. Although it is sometimes dismissed as boring drudgery, it has its place. As mentioned earlier, memorizing basic math facts, poems, and lines in a play are just a few examples. There is an important caveat to emphasize when asking students to use repetition as a tool for memory: it must be error-free repetition. For example, if the assignment is to learn a poem to be presented orally, students begin by reading the first few words, covering them over, and repeating them aloud. Only after the first few words are remembered and spoken aloud should they move on to the next few words. They should not read through the whole poem and then try to repeat the whole thing aloud. Remember, what fires together wires together; if students unintentionally repeat the wrong words, they will wire together a neural network with the wrong words.

3. Use the Same Neural Networks to Practice as to Perform. If students are studying for a written test, they should practice by writing their answers to practice questions. If, as in the poetry example, they are memorizing something to recite orally, they should speak aloud as they are practicing. This is again using the neurological adage "what fires together wires together" to their advantage. Making the writing or speaking neural networks fire together during practice will make them more likely to fire together when they are needed during a test or an oral recitation.

4. Emotions in Class Can Be Your Ally. We have reiterated the problems caused by stressful emotions, but using the resonance circuits in your relationship with your students can help them learn and remember. Showing your passion for a topic (like Mr. Arker did in his history class) or your faith in students' ability to grasp concepts (like Mr. Blanden (from Chapter 3) are examples of how positive emotions can be used effectively to help students remember. Keep in mind that some emotional expression on your part stimulates your students' limbic region to tag the information as important and memorable. Once you grasp the concept, many ideas may come to mind. For example, video clips with emotional content, using dramatic scenes from historical fiction to teach historical facts, dramatic movies, emotional musical lyrics all come to mind as possible teaching tools that use emotions to help students learn.

Chapter 6

Nurturing Student Strengths

TEACHING IS MORE THAN A PROFESSION OR A COLLECTION OF techniques—it is a way of life, a state of mind. Dan Siegel defines state of mind as "An overall way in which mental processes, such as emotions, thought patterns, memories, and behavioral planning, are brought together into a functional and cohesive whole. . . . A state of mind coordinates activity in the moment and it creates a pattern of brain activation that can become more likely in the future" (2012a, p. AI-77). An individual state of mind in regard to teaching is created over time by multiple factors including (but not limited to) training and exchanges with students and colleagues within the subtle yet powerful influence of the school culture.

This chapter considers a particular aspect of state of mind—a tendency to focus on either weakness improvement or strengths expansion with students. Usually school culture encourages the former, with a focus on test scores and the underlying assumption that all areas of study are equally important for all students. As a result, teachers have a tendency to remain alert to what is *not present* in students, and when they discover a skill or some information students do not possess, they teach it. When you train the flashlight of your attention on this alone, you tend to unknowingly dismiss or ignore what is already present in students. With a shift in perspective, you can focus on the inner strengths and talents your students already possess, collaborating with their natural gifts while helping them develop in areas of challenge. It is a simple concept that is not easy to implement.

Although you know from experience that if your students have particular areas of chronic weakness, it is unlikely that all the additional help you offer will bring them to the level of excellence in that area, but concern for their future well-being and other pressures keep you working on weakness-repair. When the challenge remains, we teachers tend to increase the level of help in ever-escalating rounds of time and attention. However, this process unintentionally creates a slowly developing problem: as the help increases and students spend more hours focused on their weaknesses, there is less time and energy for expanding their strengths.

If you step into the shoes of your students, you may be able to get a sense of the emotional toll taken by making this approach central to the teaching paradigm. From the moment the student knows that a report card with a low grade is going home, the level of tension and fear begins to rise. Because parents love their children and want them to do well, they will likely focus on the low grade(s), maybe peripherally noticing the ones that are in the acceptable range. In terms of neurobiology, this happens because the support system—parents and teachers—is experiencing fear, a state that causes all humans to adaptively focus only on information related to the perceived threat. As strengths fade into the background, students might begin to feel a whole range of emotions: shame, sadness, the sense of being different from the other kids, the pain of disappointing their parents. Each child arrives at a different combination of uncomfortable feelings depending on past experience and what their parents and teachers reflect about this situation.

As teachers know, these concerns breed urgency and a need to develop plans for improvement. If these interventions don't lead to improvement in performance and test scores, more help is brought in, and just like the student, teachers feel less and less competent. All of these unintended consequences are the opposite of what the support system intends for the student.

None of this is to say that teachers need to abandon efforts on behalf of helping students toward whatever improvement is possible. It does matter that they gain enough proficiency in areas of challenge to do well enough in the world. However, as your attention broadens to include strengths and you see students succeeding in those areas, fear and urgency about improving the weaker areas can moderate. A balance between strengths expansion and weakness improvement may be optimal for students. Let's turn to the story of how fostering strengths puts people on the road to a satisfying and successful life.

The Road Toward a Focus on Strengths

Some of the early research on supporting strengths occurred in the business world while looking at the characteristics of people doing exceptionally well. One major force behind the inception of this research was the late Don O. Clifton (Buckingham and Clifton 2001). Clifton was CEO of the Gallup Organization and intrigued by outstanding people from all walks of life—star athletes, teachers, doctors, factory workers, hotel cleaning staff, and a host of others. Whenever his organization discovered such people, he had them interviewed, seeking to understand what made them different from the rest of us. The research is ongoing and there are now a total of more than three million interview transcripts. About fifteen years ago, Gallup began mining this data treasure trove. Among other factors, they found that highly successful people discover their strengths and then put a lot of work into improving them (Buckingham and Clifton, 2001; Rath 2007; Rath and Harter 2010).

In 2012, in response to Gallup research suggesting that more than 90 percent of Americans do not believe that high school graduates are ready for college or the work world, leaders from more than twenty-five national education, business, philanthropic, and policy groups, including Gallup, came together to form the Career Readiness Partner Council (CRPC), coordinated by the National Association of State Directors of Career Technical Education Consortium. The CRPC's goal is to enhance reform efforts for college and career readiness and bring clarity and focus to what it means to be career ready, prepared "to achieve a fulfilling, financially secure, and successful career" (Hodges 2012, p. 1). Among the conclusions drawn from their research was that:

> developing one's innate talents is essential for career readiness and necessary for engaging in today's fast-paced, global economy. While academic, technical, and workplace knowledge, skills, and dispositions vary from one career to another and change over time, one's innate talents remain consistent. . . . The unprecedented level of self-awareness that results from developing one's strengths is particularly beneficial during times of transition, such as when students move from high school to ongoing education, students graduate from school and join the workforce, and employees experience a job change. (Hodges 2012, p. 1)

Although this research focuses on high school–age students, offering a strengths focus from the first day of school can only enhance and more deeply ingrain these benefits.

There is a second strand to this story that unfolds in the world of psychology. Like education's concern with weaknesses, the field of psychology has had a long-standing focus on healing pathology. For about a century, research in this field has focused on alleviating depression, anxiety, the effects of trauma, and a host of other difficulties. However, a shift began to occur approximately twenty years ago when Martin E. P. Seligman (2002) and a group of other thinkers began to realize that a person who is no longer depressed is not the same as a person who is actually happy; a person who is not anxious is different from someone who is relaxed. In other words, the lack of psychopathology is qualitatively different from the presence of well-being.

From this insight, the field of positive psychology was begun. Positive psychology focuses on what is right with people, helping them uncover the potential for states of mind that open the door to sustained well-being and resilience. Like the Gallup group, positive psychologists study successful people to understand the common factors that lead to this way of perceiving the world. Healing pathology and the support of positive states of mind are not conflicting approaches. They complement each other the same way a strength-nurturing state of mind complements a weakness-improvement state of mind.

Why Is it Difficult to Change the Direction of Our Flashlight of Attention?

Paradigm shifts are not easy under any circumstances, and the weakness-improvement to strengths-nurturing shift may be quite challenging for a number of reasons. Awareness of these potential roadblocks can make them easier to spot and step around.

- Weaknesses may be easier to measure. Most standardized testing and its interpretation, especially when used in the special education system, are designed to detect weaknesses and pathology. Standardized strength assessment is in its infancy, with most applications oriented toward adults. In the classroom it can be difficult to measure a strength that cannot somehow be translated into a standard score or a grade.
- The amygdala-centered fear system adaptively notices what is wrong more easily than what is working well. In situations you perceive to be dangerous, it becomes neurologically important to screen out extraneous information, even positive aspects, to focus on the threat so your body can prepare for fight or flight.
- When fear and anxiety are present, in addition to keeping you focused on the perceived source of threat, it is difficult to take in new information or entertain the possibility of change. Change, even the most positive change, is stressful, so when the system is already taxed, it is reluctant to add more.
- The "negativity bias" (Fredrickson 2009) is powerful. The results of numerous studies in psychology show that negative emotions are felt more intensely and have a more powerful and more long-lasting effect on people than positive emotions (Baumeister et al. 2001; Fredrickson 2009; Baumeister and Tierney 2011). This applies to your students and to you. The intensity of feelings such as fear, anger, shame, hate, discouragement, and disgust that can accompany a focus on weaknesses can quickly color your overall experience to the point that you are unable to perceive anything potentially positive about a situation or person, making it very difficult to focus on strengths. This chapter explores a way to navigate this particular impasse a bit later.

How Can We Change Our Focus?

Now that you have some idea of the obstacles, what might you do to increase your awareness of your own and your students' strengths? One place to begin is to listen for something called *explanatory style*. When you have had a difficult experience, you tell yourself a story about what it means. For example, if I have a mental model based on implicit memories of being a disappointment to others, my story of a difficult experience is likely to include that theme (more examples will follow).

Research tells us that how people view the cause of events has profound consequences for mood, learning, and physical health (Seligman 1990, 1995, 2002). This attribution of causes is called an explanatory style and there are two varieties: *optimistic* and *pessimistic* (Seligman 1990, 1995). Before the specifics, let's consider what optimism and pessimism are. The cliché "optimists see the glass half full and pessimists see the glass half empty" does not adequately capture the difference that shows up in research. Rather than the event itself, the difference lies in how one views the *cause* of the event. In other words, it is not whether the glass is half full or half empty, but how it got that way. Optimism is not the same as being happy and looking on the bright side. It is a belief about causality. Optimism distinguishes itself from pessimism along three dimensions: permanency, pervasiveness, and changeability. Seligman offers the following explanation. "When a negative event occurs, an optimist believes, 'It's going away quickly, I can do something about it, and it's just one situation.' A pessimist believes: 'It's going to last forever, it's going to undermine everything, and there is nothing I can do about it'" (2011, p. 189).

Let's apply this to an unpleasant event for some students: failing a math test. One student might think: "I'm so disappointed I got an F on the math test. I always do poorly on math tests. I'm stupid in math." These statements reveal the student has a pessimistic explanatory style. Using Seligman's criteria:

- It's going to last forever ("I *always* do poorly on math tests"; implies a permanent condition regarding math tests.).
- It's going to undermine everything ("I'm stupid in math"; not just this test, I'm stupid in all math).
- There is nothing I can do about it ("I'm stupid in math"; obviously, there is nothing I can do about being stupid!).

A student with an optimistic explanatory style might think: "I'm so disappointed I got an F on the math test. I did a lousy job. I didn't study." It is important to notice that a student with an optimistic explanatory style does not avoid the feelings of disappointment nor replace them with happy thoughts. The optimistic student feels just as disappointed as the pessimistic student, but believes differently about the cause. Again, from Seligman:

- It's going away quickly ("I did a lousy job"; this implies a temporary situation).
- I can do something about it ("I didn't study" suggests that the situation can be changed in the future with more studying).
- It's just one situation. ("I got an F on the math test" confines it to this single experience).

Initially, it can be surprising to imagine that it can be a sign of optimism for a student to say, "I didn't study." However, in this case it certainly is, because it implies a temporary condition that can be corrected. Students who believe they failed a test because they are inherently stupid have a sense of helplessness and will be harder to encourage toward more effort than those who perceive the test result had something to do with their action (or lack of action). Pessimistic explanatory styles can be deeply rooted in implicit memory because of numerous experiences that create a negative view of character ("I'm lazy"), capacities ("I'm stupid"), or support ("no one will help me"). Shifting the resulting helplessness and broad-based conviction that there's nothing to be done requires continuous effort on our part to gradually provide enough disconfirmation that a new implicit root is planted and flourishes.

Here are some more examples so we can cultivate our ability to hear the difference between explanatory styles.

Permanent versus Temporary
Pessimistic: "I hate Ms. Smith; she's always grumpy and yells." "I can't make friends in this new school."
Optimistic: "I hate it when Ms. Smith is in a grumpy mood and yells." "It takes time to make friends in a new school."

Pervasive versus Specific
Pessimistic: "I'm not a good athlete." "I'm a terrible math student."
Optimistic: "I'm not good at basketball." "I'm terrible at long division."

Changeable versus Unchangeable
Pessimistic: "I got a D, because I'm stupid." "I'm lazy."
Optimistic: "I got a D because I didn't study enough." "I didn't work very hard."

It is always a good idea to begin with yourself. One way to do this is to ask a fellow teacher to partner with you and listen for times of both optimistic and pessimistic styles. It is very challenging to hear yourself because you've likely been doing whatever you're doing for so long you've become deaf to its meaning. In other words, you have unintentionally wired together a network of neurons by repetitively using an explanatory style. It is also likely that you are optimistic in some areas and pessimistic in others. Tracking the feelings in your body when you're in either mode can help you get a felt sense of what it is like for your students. In addition, with practice, those very bodily sensations will alert you when your students are using one style or the other.

As you become more skilled at listening, when you hear students use a pessimistic explanatory style, you can create a disconfirming experience by offering an authentic and positive offsetting observation about something you have noticed: "I don't see you as lazy. You just don't work as hard on subjects you dislike. I have seen you work hard on things you love. I remember the time you worked really hard on ________." Helping students spot and change their explanatory style can be an incentive to notice strengths so you will have a reservoir of positive attributes and experiences to offer whenever a pessimistic explanation surfaces. Understanding the depth of the roots of these explanations can give you the patience to repeat this experience as many times as necessary.

It is important to teach and model an optimistic explanatory style because research is beginning to show that this has consequences on physical health. In Seligman's book *Flourish*, he notes that science is beginning to show "optimism is robustly associated with cardiovascular health and pessimism with cardiovascular risk . . . positive mood is associated with protection from colds and flu and negative mood with greater risk for colds and flu. . . . Highly optimistic people *may* have a lower risk for developing cancer . . . healthy people who have good psychological well-being are at less risk for death from all causes" (2011, p. 204). It is never too early or too late to support development of an optimistic viewpoint.

Strengths Spotting

Hand in hand with working on your awareness of explanatory styles, you can be aiming your flashlight of attention on student strengths. You might begin by acknowledging that a strength with car engine repair or a passion for gardening can be just as important as academic achievement to expanding a student's knowledge. Learned beliefs about education may make this seem like heresy until you bring your car in for service or need to shop for fresh vegetables at the farmer's market. Helping students begin and continue the journey of discovering and expanding their strengths is important no matter where it leads them.

To help you spot your own and student strengths, here are a few things you may notice (there is a more detailed list in the Tools for School section at the end of the chapter). When you speak about subjects in your area of strength, you will have a sense of energy and uplift in your voice and seem relaxed yet energized. The conversation will be free flowing and the descriptions rich; you will have a sense of absorption in the topic (Linley 2008). In the chapter on attention, we talked about the state of flow, which occurs when our strengths match the demands of the task. Enlisting students' strengths at school creates ease of learning, enthusiasm, delight, and increasing skill. What student wouldn't be eager to come to school if he or she could count on being met and supported in these kinds of explorations?

As you spot strengths—your own, your students'—it may be helpful to keep a written note of them in a particular place, so even in the pressure of the school year, even at times when you must concentrate on repairing weaknesses, you can review them and, through repetition, begin to see these strengths whenever that person comes to mind.

Strength Teams

If you are teaching at the middle or high school level, you may not realize that you already have a strength expansion program in school—sports teams. Most sports teams select students with athletic strengths through a series of tryouts that enable coaches to choose the most talented students. The students picked for the team improve their already strong skills with practices, drills, and games. Observing student athletes can give you a sense of the motivation, energy, and commitment that is possible when students work to improve their strengths.

Similar experiences of strength expansion often happen in extracurricular activities, like the debate team, the drama club, band, or the chess club. In these settings, people's strengths are emphasized and other team members make up for their weaknesses. A friend of mine was too shy to appear on stage (a weakness) but loved the theater environment (her passion), so she used her organizational strength in the role of stage manager to the benefit of all. Most Major League Baseball pitchers can't hit or run all that well, and no one seems to mind as long as they win games with their hurling arm. How might you use your creative imagination to foster such teams in your classroom? Each situation will be different, and making a map of student strengths can begin to point the way.

It is true that you may already have untapped strength teams in your classroom. Consider the following all-too-familiar scenario. You've planned your lesson around the use of some form of technology, and the technology won't work. As your frustration and embarrassment mount, students spontaneously spring into action. Some are helpful and others not so much. The frustrating and embarrassing situation can get out of control as wires are plugged in and unplugged and things are clicked on the computer. This scenario can be a teachable moment and an opportunity to appoint a classroom technology strength team. While the students are trying to fix the technological glitch, talk to them about their technological strengths. Ask if they would like to be on the classroom tech team. Students with strengths in this area, if publicly designated as a team, will probably rise to the occasion. Some may study independently the specific technology used in the classroom to show off their skills the next time there is a problem. If you are especially fortunate, your technology team will include students who do not shine in your classes otherwise. Highlighting their technology strengths will give them and the whole class a better perspective on the balance between strengths and weaknesses.

Peer tutoring is another way to create a strength team. For example, if John needs help with writing a report and Bill has a writing strength, instead of just asking Bill to help John, first describe John's strengths (academic and nonacademic) and Bill's strengths to both boys. Frame the whole process as a team designed to have John turn out a good report, and maybe offer both boys extra credit if the report is excellent. The strengths team approach puts both boys on the same team focused on improving the report, creating a balance. Students who need help benefit from hearing the teacher describe their strengths, and tutors benefit by hearing you describe their strengths.

Your Home (Strength) Team

One resource for creating strengths teams in the classroom is to begin to explore the ones you already have in place in your personal life. Many adults have moved away from their weaknesses and toward their strengths through a series of small decisions that led them to successful and satisfying careers. They have unwittingly developed strength teams with their friends and intimate partners. To see if this is true for you, pause here, reflect on your life, and answer these questions:

- What are your areas of weakness?
- How are people helping you work around these weaknesses? For example, if you have a math weakness, how do you now handle your checkbook and income taxes? Does your spouse do it? Do you hire a tax preparation service? Do you clumsily do the finances and wish you could do a better job?
- How about other areas that might be weaknesses: Cooking? House cleaning? Repairing things around the house? Have you found ways to deal with these or other common tasks? Does your partner do most of the cooking or house repairs? Do you just let the housekeeping go, and feel guilty about it? Do you wish you could hire someone to do it?

Like the rest of the human race, you have weaknesses and may have unknowingly developed ways to work around them with the help of others, just as you may be unknowingly using your strengths to help others work around their weaknesses. These examples from your home life can help you see opportunities for the same approach in the classroom.

Strengths Assessments

A number of people and organizations have developed valuable formal assessments for identifying strengths. The three instruments summarized here are available online, and there are details as well as links for all three in Tools for School at the end of this chapter.

Strengths express themselves in numerous ways. Christopher Peterson and Marty Seligman (2004) define *strengths of character* as moral traits that can be expanded and built on regardless of the context. They have developed a taxonomy of character strengths purposely paralleling the model of the Linnaean classification of species. It has three levels. At the most general, there are six virtues: wisdom, courage, humanity, justice, temperance, and transcendence. These broad categories are endorsed by all of the major philosophies and religious traditions. In other words, people everywhere generally agree that they are desirable.

Under each of these six virtues, Peterson and Seligman enumerate the second level, character strengths. Character strengths are the ways an individual expresses a virtue. For example, the virtue of wisdom may be expressed through one or several of the character strengths: curiosity, love of learning, creativity, and/or perspective taking. There are a total of twenty-four character strengths. The assessment they developed rank orders your twenty-four character strengths emphasizing your top five. The third and most specific level of this taxonomy is situational themes, which seek to add context. For example, a child with the character strength of curiosity might express it differently in a school setting than she would in a home setting. This level of the taxonomy attempts to account for these differences.

The Gallup Organization has also developed an assessment instrument for adults, which might be ideal for beginning to understand our own strengths. Their research with highly successful adults noted earlier showed approximately 400 common themes, and they distilled these down to 34 and built the assessment around those. There is also a version for use in higher education called Strengths Quest.

A third strengths assessment was developed in the United Kingdom and is based on a model that considers an individual's strengths on four dimensions: realized strengths—high energy, high performance, and high use; unrealized strengths—high energy, high performance, but low usage; weaknesses—low energy and low performance; and learned behavior—low energy and high performance. This final category is interesting because it accounts for tasks we may have learned to do well but find emotionally draining (Linley 2008).

It is important to remember that the results of these assessments are just the beginning of the strengths expansion journey. The strength labeling only gives a focus for the next step of applying those strengths in our daily lives.

Beginning with Ourselves

When a school wants to shift its culture in the direction of support for strengths, a first valuable step is for all members of the faculty and administration to use one of the assessment tools to uncover their gifts. When discussing their results, weaknesses will undoubtedly come into the conversation. It helps to focus the discussion on how the school can use this awareness so that one person's weakness is balanced by another's strength. This is an exercise in wisdom and humility that can move a school down the road toward appreciating more fully both the beauty of students' strengths while granting the sense that their weaknesses may not be as detrimental to their future as one might imagine. If you are able to take your experience with yourself into the classroom, you may be able to create teams in which students have the invaluable felt sense experience of their weaknesses being made up for by another, and their strengths being used more fully. A strength expansion–weakness improvement balance is an implicit mental model worth cultivating.

A concrete example may help you imagine doing this. A local school decided to experiment with exploring a strengths-based focus, mostly because everyone was worn out from a steady diet of weakness correction. They began their journey with strengths expansion when all of the administration, faculty, and staff completed two strengths assessments. They implemented their growing awareness of their strengths in the day-to-day running of the school. When tasks needed completing or committees needed to be created, they did so based on the strengths found through the assessments. It was quite a surprise when the school director's results showed that his top strength was Maximizer—"people especially talented in the Maximizer theme focus on strengths as a way to stimulate personal and group excellence. They seek to transform something especially talented into something superb" (StrengthsQuest 2000). The act of labeling his strength focused it and activated the rest of the staff. Years later he continues to state it bluntly: "I'm a Maximizer. We are doing well. We can do better." By applying it to himself first and also openly and regularly acknowledging his weaknesses, his approach seems to have added an optimum amount of pressure on the whole school because teacher performance improved (and continues to improve).

At a middle school, several years ago the administration decided to create a team of two co-principals from members of the current staff. Administration, faculty, and staff together created a master list of all the duties that needed to be managed. Using their strengths, the two prospective principals spent several hours choosing their duties from the master list. Not surprisingly, there were duties that neither wanted. They negotiated who would take the unwanted tasks until all were delegated. They then brought their choices back to the staff and discussed how their strengths and weaknesses led to their choices. They also spoke of the duties that neither wanted but chose anyway. The discussion led the teaching staff to respect the complexities and stresses of each principal's position. It took longer for everyone to learn who was responsible for what than if they had simply divided the positions by grade level, but this was a small price to pay for the success that followed. The openness of the process even brought a positive perspective to the things no one wanted to do. Later, it led to good-natured humor among the staff when one or both principals neglected some of the unwanted tasks. As the process evolved, some staff members stepped up and, based on their own strengths, took on some of the unwanted tasks.

Allen Discovers his Strength: A Case Example

Sometimes, it's the students who begin to teach the staff about how a strengths-based focus can be transformative. When Allen started at my school in the seventh grade, he was reading at about the second-grade level. His mother was frantic and his teachers from his previous school appeared resigned to the fact that Allen would never learn to read very well. He was promoted year after year and received extra help with reading, but with only minimal improvement.

When Allen first came to the school, he appeared small for his age, socially awkward, and downtrodden. The staff would see him quietly walking through the halls, head down, making no eye contact, and looking at his shoes when he spoke. Although his previous school seemed to have given up on his reading, his mother hadn't. She pushed to have him placed in this new school, and on his admission, she demanded that he have two hours of daily one-to-one reading instruction. He also received reading help in his English and social studies classes. His whole school day was focused on improving his significant reading weakness. At home his mother read to him and he read to her almost every night. Near the end of his first year at the school, he was tested and found to have made absolutely no progress. In fact, the reading tests indicated that his achievement may have dropped slightly.

His mother and the special education coordinator from his previous school were irate. Allen was at the meeting, and as the testing results were described, he began to cry. The special education director of the school district angrily explained that he was being sent to this particular school only because of his low reading achievement, and if the teachers here could not improve his reading he would be removed from the school.

As the school year ended, Allen appeared even more downtrodden, if that was possible. He left for summer vacation feeling defeated. No matter how hard he worked, no matter how much instruction time he received, no matter how many lectures he got from his parents, his reading had not improved. The sad thing is that the singular focus on his weakness had unintentionally taught him a pessimistic explanatory style. He felt helpless to change anything. His pessimistic explanatory style may have isolated him from other students, because he had not made many friends that first year. As he left for summer vacation, it appeared that he might simply quit.

In spite of the initial focus on his weakness, Allen was able to metaphorically jump out of the box. During the break, he joined the school's summer program and became actively involved in the outdoor activities, along with his reading tutoring. This gave him the opportunity to connect with other students and staff who shared his interest. His athletic abilities began to emerge as a strength. As the next school year began, he joined the soccer team, often hiked during the school's experiential learning days, and improved his skiing in the winter. His strengths had begun to blossom. His daily two hours of individual reading instruction continued, and Allen's physical activity increased. His interpersonal strengths began to emerge on the sports teams and he developed positive relationships with coaches and peers. He was becoming well liked at school. As the year moved on, his head rose as he walked through the halls, and he looked up when he spoke. He grew several inches and turned from a downtrodden boy to one with positive close relationships with all of the staff. His mother became active at the school, helping with many events.

As the emphasis on his strengths continued, Allen became an important member of the school culture. As the spring of his second year arrived, so did the time for the inevitable annual reading test. Allen, his parents, and the school staff were all worried about the results. To everyone's astonishment and delight, his results showed that his reading achievement level had risen three years in the previous eight months. It was an amazing amount of growth that was celebrated by his parents, teachers, and even the special education director of his previous school.

As of this writing, two years after those testing results, Allen is reading at about the eighth-grade level, is a member of the snowboarding team at a local ski resort, has discovered he is also strong with hands-on projects, and spends weekends making things in the family's workshop. This previously downtrodden boy is also the recipient of a good deal of teenage female attention.

When a student is about six years behind in reading, it certainly grabs educators' attention and demands an intensive instructional focus. Allen received the instruction he needed from a gifted reading teacher, but it was not enough. In hindsight everyone realized that his dramatic deficit grabbed attention to the exclusion of everything else (teachers dedicated to improving student performance could speculate that it had elements of nonconscious bottom-up attentional control). For Allen to improve his reading, he also needed time to recognize and develop his strengths in connection with his peers. Educators needed to intentionally use the top-down attentional control to expand their perception and place him in situations where he might discover and deploy his strengths (e.g., outdoor summer program, athletic teams) even though they seemed to have no direct link to reading instruction. Someone with a narrow reading improvement focus might argue that a membership on a soccer team, snowboarding, and a summer program with anything other than reading instruction were a waste of time and resources, but this was not the case for Allen.

The shift from the narrowly focused weakness-elimination strategy to a strengths-nurturing strategy balanced the instruction for the weakness and not only helped Allen improve his reading, it fundamentally changed his developmental trajectory.

The Positivity Offset

Earlier, I talked about the negativity bias and the power of painful emotions to color your world and shape your attitudes and actions. As you shift toward a strengths focus, the "positivity offset" explained by Barbara Fredrickson (2009) will likely come into play. Here is how it works. Research shows that negative emotions are felt more intensely, but positive emotions are experienced more often. The tipping point when negative emotions are offset by positive emotions is a ratio of about 3:1. The positivity offset is at least three positive emotions to the experience of one negative emotion. The effect of this 3:1 practice has been demonstrated in other settings. For example, flourishing marriages that last have a 5:1 positive to negative ratio and those that end have a 1:1 ratio (Gottman and Silver 1999).

Sadly, the general population has about a 2:1 ratio. In a study about depression, individuals suffering in this way have about a 0.5:1 ratio and after treatment return to approximately a 2:1 ratio. If we cultivate a state of mind that notices, acknowledges, and improves student strengths, we can draw closer to the 3:1 ratio (Fredrickson 2009).

A shift to focusing on teacher and student strengths makes it seems likely that all will move both internally and outwardly into the territory of 3:1 (at least). At the same time, students will begin to feel seen for who they are at their best. This is certain to make the classroom feel like a safer place, laying the foundation for secure, warm, respectful relationships as the gateway to enhanced learning.

Let me offer a final piece of research that points the way toward student success, again from the Gallup Organization. "The best educators know that for students to achieve meaningful, lasting success in the classroom and beyond, they must be emotionally engaged in the educational experience. This means educators must focus on students' hope, engagement, and wellbeing—the predictors Gallup has discovered matter the most (Gallup n.d., p. 1). Nationally, about 50 percent of our students report feeling hopeful, while the other 50 percent are stuck or discouraged (Lopez 2013), a statistic that roughly parallels attachment research on security and insecurity. Although no one can know if that correlation holds exactly, what the research indicates is that engagement with teachers is a crucial factor in cultivating the fertile soil for success, more important than native intelligence or socioeconomic status. An approach that takes into account the primary importance of connection and implements a strengths focus has an excellent chance of using engagement to build hope that leads toward well-being.

Tools for School

There are numerous ways to develop a strengths expansion model in your school or classroom.

1. Strengths Expansion Begins with You. To fully understand the power of a strength expansion approach, you need to experience it yourself, and so will the administrators and teachers in your school. Take one of the online assessments (websites appear in the next item) and use strengths spotting (details in number 4) as a team at school. Put what you learn by these means to work in your personal life and whenever there is a decision to be made about who will do which tasks.

2. Strength Assessments. Here are three assessment tools to help you, the staff at your school, and your students discover their strengths.

- VIA Signature Strength Survey. You can identify your five top character strengths out of the list of twenty-four by taking the VIA Signature Strength Survey (http://www.viacharacter.org or http://www.authentichappiness.sas.upenn.edu). It is free and a good place to begin the exploration and expansion of strengths of character. For a fee, an in depth individual and a team report is available at http://www.viacharacter.org. At the authentichappiness website, there is a shorter version for younger children.
- StrengthsQuest or StrengthsFinder. Developed by Gallup (http://www.strengthsquest.com), this is a useful tool to begin a journey of discovering and expanding strengths. It is organized around thirty-four themes; there is a fee for taking this instrument.
- Realise2. A third assessment is the Realise2 (http://www.

cappeu.com). It offers a balance between strengths and weakness identification. There is a fee for this service.

3. Strength Teams. Highly successful adults spend a good deal of time and energy on improving their strengths while working around their weaknesses. Work with your fellow teachers and students to develop creative ways to implement strengths expansion and weakness workarounds. Create teams in your school, classroom, and personal life using others' strengths to balance your weaknesses and vice versa.

4. Strength Spotting is an observational tool you can use with your students, fellow teachers, and yourself. Alex Linley has created a list of clues to help you spot strengths and I recommend either of his books to help you further (Linley 2008; Linley, Willars, and Biswas-Diener 2010;). Here is a modified version of his list.

- Energy: When someone is using a strength, it can feel energizing. Look for activities that take effort but leave you more energized when you've completed them than before you started.
- Voice Tone: An increase in energy and animation as someone describes an activity is a clue they are describing an area of strength.
- Vocabulary: When you hear yourself or others use words such as "I love to . . ." or "It feels great when I . . ." they are likely talking about a strengths-based experience.
- Rapid Learning: When students learn something rapidly and effortlessly, it may be a sign they are learning in an area of strength.
- Ease: If something is surprisingly easy to complete, it may be a clue that a strength is being used.

- Attention: When things easily draw attention, a strength is likely in the picture.
- Flow: Listen for statements such as "I thought I was doing ______ for fifteen minutes, but it was really two hours! Where did the time go?"
- Motivation: What things do you do just for the joy of doing it? You would do it even if there were no external motivators such as pay or threat of punishment.
- "To do" list: This is a personal favorite strength-spotting clue of mine. What do you always get done without putting it on a "to do" list? What things do you never have to be asked to do?

5. What Went Well (WWW). A simple tool for school you can use at the beginning or end of your classes is the what went well (WWW) exercise. This activity is among the most thoroughly researched exercises in positive psychology (Seligman 2002, 2011). It is described in Chapter 4; here are some specific instructions: To implement this approach, begin each class with the question: "What went well (WWW) for you since our last class?" Or "What are you grateful for since our last class?" When you first ask the question, be prepared to observe the effect of the negativity bias. Students will have experienced negative events more intensely than positive ones and believe that they are more important to share. At first the WWW question may elicit responses like "Nothing," "I don't know," as well as a chorus of eye rolls, especially if you are a middle or high school teacher. All you need to get this started is a few students and your own WWW. As the routine develops, you might want to call on certain students to help them change their perspective.

6. Standardized Testing Reports. If you are a school psychologist or special education teacher and are called on to test students and write interpretive reports, enumerate student strengths to the same extent that you identify their weaknesses in your report. Be sure to include recommendations to expand areas of strength as well as the usual strategies to improve weaknesses. Entry into the special education system is based on discovering student weaknesses, so testing reports necessarily emphasize them. Students, however, are more than their weaknesses, and expanding on their strengths offers the student, parent, and school team a more complete picture of the student than a singular focus on weaknesses and learning challengers.

7. Storytelling and Rituals. Annual celebrations in school offer an opportunity to spotlight student strengths. Storytelling is an effective tool for highlighting strengths. Jennifer Fox Eades's book *Celebrating Strengths: Building Strength-based Schools* (2008) is a valuable resource, especially for elementary schools.

8. Positive Psychology Activities. For an adaptation of many positive psychology exercises to a school setting try *SMART Strengths: Building Character, Resilience and Relationships in Youth* by John Yeager, Sherri Fisher, and David Shearon (2011). This book includes fifty pages of activities you can adapt to teaching at the middle and high school level.

Chapter 7

Mindfulness: The Mind and Brain Collaborating

THERE ARE STILL TOO MANY DAYS LEFT BEFORE THE WINTER BREAK. the weather has been cold, cloudy and dark for far too long. Beleaguered staff and students sit together in self-imposed silence in a large room, waiting for another school day to begin. The staff's winter exhaustion is as persistent as the students' winter restlessness. We nonetheless begin the day with a few minutes of mindfulness. Walking slowly down the center of the room, I match my pace to the low energy. I begin talking to this group of oppositional teenagers. "You have the right to control your own emotions. Why should you let this lousy weather control you and make you feel miserable? For that matter, why should you let anyone make you feel anything?" I now have the attention of many students and staff.

I stop walking and talking, pausing long enough for the inattentive ones to look up with questioning eyes. "I'm not telling you to cheer up. I'm not going to tell you how to feel. It has been cold, cloudy, and miserable for days and it is a long time until vacation.

You can let the weather determine your mood, if you want, but let's just take two minutes, look inside, and observe how we are feeling. Don't try to change it—just observe. Let's start like we usually do by paying attention to our breathing, then pay attention to how we feel. You might begin to realize there is you, and there are the feelings, and they are different." I ring a small bell that was given to the school by an enthusiastic parent who found that the mindful minute helped her son (who has bipolar disorder) calm at home. I keep one eye on my watch and the other on the students, all sitting quietly, most with their eyes closed. I ring it again to end the mindful minutes. The energy in the room has shifted, from an isolated, restless exhaustion to curiosity, connection, and sharing. Students raise their hands for daily announcements, which includes one reporting the previous day's sports scores, another a current event he found interesting, and a teacher sharing a personal story from the evening before. The announcements complete, students and staff leave the room with a good deal more positive energy and interpersonal connection than appeared possible before the two minutes of mindfulness.

What Happened in Those Two Minutes?

The foregoing is an example of a brief mindfulness practice that can be used at school. It enabled staff and students to internally step back from feeling tired and miserable to be present with themselves and each other in the moment. Because the students have been practicing several times a day all year (and the staff for many years), the experience of quiet inner focus has become very effective at allowing them to separate their sense of self from the flow of emotions that constantly move through them like a river. Repetition has strengthened their internal observer's capacity, so it became a reliable way to help everyone settle into connection, first internally and then externally. As you will see, actual changes in brain function and structure can put you in touch with the internal resources that allow for both warm relating and a settled sense of well-being. Coming to the present moment creates a sense of newness that refreshes and connects. When these practices are woven into the flow of daily school life, students begin to request the mindful moment even if the teacher forgets.

What Is Mindfulness?

Let's start by considering your typical internal state. As your mind jumps from one thought to another, you pay attention to the outside world just enough to engage in the required activity before you. Some research shows that even aside from daydreaming, no one attends to what is directly before them more than about 50 percent of the time (Jha 2013). Perhaps you have noticed this state of mindlessness when sitting in a faculty meeting mentally making a grocery list; when you're reading, and arrive at the end of the page having no idea what you just read; or when listening to a lecture but thinking about vacation. Instead of being present in your life at this moment, your mind has been engaged in mental time travel into the future or the past. Mindlessness may be a side effect of human evolution. As discussed in Chapter 4, your attention is pulled by anything unusual because it might pose a danger. When daily life and the present moment are the usual humdrum, little effort is expended attending to it because you save your attentional resources for the novel and potentially dangerous.

For example, what details of your most recent morning commute to school can you remember? If it was uneventful, then you probably have difficulty distinguishing which details you remember happened on which day. It is as if your car drove itself on automatic pilot. If you were teaching a lesson you have taught previously, it may be that you taught in a similar mindless state. Humans function well enough without focus because we are able to have the automatic pilot, the habitual responses to the usual stimuli, direct our actions. It is a useful capacity in its own way. It means that we don't have to learn anew how to do many patterns of action, but instead use these shortcut networks we have developed through repetition to mindlessly and effortlessly glide through the day. However, in this mindless state, we miss what is uniquely happening in this particular moment, one of a number of significant drawbacks. For example, one of the most in-depth analyses of car accidents, done back in 1979, found that inattentive driving accounts for 56 percent of car accidents, and this was well before cell phones and texting (Wang, Knipling, and Goodman, 1996). Because you have well-developed neural patterns for the basic habits of driving, you function on automatic pilot, but when the unexpected happens, you aren't able to adjust quickly enough to meet the novel situation. That is one major contrast between a mindless state and mindful one—being sufficiently present in *this* moment to engage with it as it is happening. The ability to do this would clearly enhance students' ability to learn as well.

While a secular practice, mindfulness is a state of mind that in one form or another has been defined and practiced by most of the major world religions for thousands of years (Armstrong 1993; Siegel 2007). During the past thirty years, the popularity of mindfulness outside of religious traditions has increased and scientists have developed tools to study the changes in the brain that it brings. To provide a foundation for doing this, they developed an operational definition independent of religious traditions. Their various definitions range from the complex—"attentive and nonjudgmental metacognitive monitoring of moment-by-moment cognition, emotion, perception, and sensation without fixation on thoughts of the past or the future" (Garland and Fredrickson 2013, p. 46)—to the more accessible—"awareness of present-moment experience, with intention and purpose, without grasping on to judgments" (Siegel 2012a, p. AI-51)—to the simple—"awareness of the present moment with acceptance" (R. Siegel 2010, p. 11).

As the practice of mindfulness meets modern scientific research and application, some might still confuse its current use in education with religion. There may be concerns that mindfulness is a religious practice being imported into schools. To understand how something practiced within the context of faith can be applied without any connection to religion, consider kindness. While being kind to our neighbors and the downtrodden is a part of the major religions, it is not solely a religious practice. Kindness, like mindfulness, can be and is practiced in a vast number of different settings and situations that have nothing to do with religion. Also modern research has shown the effectiveness of kindness in increasing well-being (Fredrickson 2009). As neuroscience offers a clear understanding of how the practices of inner focus strengthen the parts of the brain that are responsible for allowing students to learn efficiently and deeply, you now have some information to meet the religious concerns with empathy and knowledge.

Mindfulness can be seen as having three aspects. The first is the *practice of mindfulness*, which involves any of a number of mental and physical exercises ranging from sitting in silent meditation (cultivated in Buddhism, Hinduism, Christianity, and other traditions), focusing on energy as in qigong (Chinese Taoist tradition), and focusing attention on physical postures in yoga (Hindu tradition). Other practices share the common theme of purposely placing attention on one thing while acknowledging, observing, and letting go of distracting thoughts and emotions. What they have in common is increasing the capacity to internally direct your attention; step back and observe your thoughts, feelings, and sensations; and let them go rather than getting caught up in them.

Mindfulness practices gradually change the brain in ways that allow a practitioner to experience the second aspect—the *state of mindfulness*. In this state, we can be aware of the present moment with acceptance, and this opens the door to attending to the uniqueness of each moment.

Frequently creating a state of mindfulness by repetitively engaging in these practices can strengthen and stabilize these changes in the brain, enabling you to experience an ongoing mindful state even when not engaged in a particular practice. You might then be referred to as having the third aspect—the *trait of mindfulness*. Any of the traits or mastered skills you can observe in yourself have been similarly cultivated through repetition. Neurologically the process is probably similar to repeatedly practicing a musical instrument until you reach mastery. People who master a musical instrument play it any time "without thinking"—that is, they do not have to consciously think about finger position or note selection; they simply play.

Mindfulness Research

So powerful is this tool for creating states of focus and well-being that at the time of this writing, mindfulness is the object of research and application in more than 250 medical centers worldwide (Jha 2013). What follows is a selection of research results on mindfulness, mostly conducted with adult subjects. Some of the research used self-report measures, which are not all that rigorous but can be used to give clarity about how mindfulness influences a person's experience in real-life situations. Some of the research used fMRIs (functional magnetic resonance images), EEGs (electroencephalograms), and computer-based rapid presentation of stimuli, which are more rigorous, but removed from real life. Self-report scales tell us what people experience with mindfulness, while fMRIs and EEGs tell us what is happening inside their brains.

Jon Kabat-Zinn of the University of Massachusetts Medical School is one of the pioneers of mindfulness research. In the 1970s, he began using these practices to help patients with chronic pain. He noticed that his patients' pain lessened, but as you might imagine, the idea that mindfulness could lessen pain was considered outside of the realm of possibility and typical science in the early 1970s. To study his observations, he developed a standardized mindfulness training program, Mindfulness Based Stress Reduction (MBSR), a structured eight-week training program with two twenty-minute daily meditation sessions at home. Because it could be replicated, MBSR became the standard experimental technique used to study mindfulness in adults (Kabat-Zinn 1982).

One of the earliest researchers to use fMRI and EEG to study the brains of mindful meditators was Richard Davidson. He and his team at the University of Wisconsin are probably most well known for their studies of Buddhist monks recruited by the Dalai Lama. The monks had accumulated more than 10,000 hours of meditation in their lifetimes. The research findings were tantalizing because they demonstrated that the brains of long-term meditators showed heightened activity and a thickening in the left frontal cortex, an area known for involvement in feelings of well-being (Davidson, et al. 2003). However, 10,000 hours of meditation certainly put any broad application of the research out of reach for most people. The team began to wonder whether less time meditating could still cause the brain structure changes seen in the monks. They and others studied people's brains before and after the MBSR training and found detectable changes in the brain that paralleled their findings with the monks (Davidson et al. 2003; Lazar et al. 2005; Davidson and Lutz 2008). With prolonged practice, the changes were more substantial, but the same areas of the cortex were connecting and thickening even with a short-term mindfulness practice.

Once it was demonstrated that one did not have to become a monk to have changes in neural connectivity, the research expanded and continues to demonstrate the improvement in well-being of mindfulness for adults, including teachers. For example, a regular mindfulness practice, usually twenty minutes practice twice a day (the MBSR protocol), helps lessen burnout for teachers (Jha 2013). Mindfulness may not change the actual causes of the job stress that leads to teacher burnout, but it does change teachers' perception of the stressors and renews a sense of curiosity and connection with students and parents (Jha 2013). Mindfulness helps improve mood and protects adults from the negative effects of high stress, a finding applicable to the high-stress job of teaching (Jha et al. 2010). It has been found to alleviate depression and lessen feelings of loneliness and associated physical inflammation in people between the ages of fifty-five and eighty-five (Creswell et al. 2012). As research continues, the benefits of mindfulness for teachers and other adults is becoming clear. If you would like to experiment with some basic mindfulness techniques, there are some descriptions and suggested reading for further exploration in the Tools for School section at the end of this chapter.

Research on Mindfulness in School

Although there are well over 800 studies on mindfulness with adults, fewer studies have focused on children and adolescents in school (Black, Milam, and Sussman 2009). The preliminary research on school-based mindfulness interventions has paralleled the research findings for adults. Thus far school-based mindfulness research has shown:

- Practice led to improved emotional and behavioral self-regulation, frustration tolerance, and self-control (Wisner, Jones, and Gwin 2010).
- Mindfulness practices may facilitate increased emotional intelligence (Goleman 2011)
- Adolescents showed enhanced ability to pay attention, improved concentration, and decreased anxiety (Beauchemin, Hutchins, and Patterson 2008).
- Of seven studies examining changes in anxiety, five revealed significant decreases in anxiety that were attributed to meditation (Black, Milan, and Sussman 2009).
- Shortened periods of mindfulness practice for adolescents (ten to twelve minutes once a day) may have a similar effect as the standard twenty-minute twice a day practice for adults (Wisner, Jones, and Gwin 2010).
- Mindfulness practices appeared beneficial in different school settings, for example, public schools, private schools, and alternative schools (Wisner, Jones, and Gwin 2010).
- Research on students with learning disabilities showed that mindfulness may lower anxiety and improve social skills (Beauchemin, Hutchins, and Patterson 2008).
- Boys diagnosed with ADHD showed decreased distractibility, increased selective attention, and reduced impulsivity compared to a control group (Black, Milan, and Sussman 2009).

In regard to this last finding, mindfulness may be a method to teach your students how to pay attention, including students with ADHD (Beauchemin, Hutchins, and Patterson 2008). One kind of mindfulness practice is to focus on one thing, such as the breath, an object, or a body sensation, and when the mind wanders bring attention back to the main focus. There are innumerable external and internal distractions for this deceptively simple task, and repetitively, intentionally reorienting your attention away from the distractions back to the object of focus seems to strengthen the ability to control attention. This strengthened control of attention is useful at other times, when the child, teenager, or adult is not involved in a mindfulness practice (Napolia, Krech, and Holley, 2005; Jha, Krompinger, and Baime, 2007; Jha 2013).

A review of the research literature on school-based mindfulness interventions concluded: "The benefits of meditation interventions for adolescents include improved cognitive functioning; increased self-esteem; improvements in emotional self-regulation, self-control, and emotional intelligence; increased feelings of well-being; reductions in behavioral problems; decreased anxiety; decreases in blood pressure and heart rate; improvements in sleep behavior; increased internal locus of control; and improved school climate" (Wisner, Jones and Gwin 2010, p. 152). This is an impressive list of benefits, but it is important to note that research in this area is still in its infancy. Researchers continue to note the methodological weaknesses of some of the studies, stressing the continuing need for rigorous research on mindfulness with children and adolescents in school settings. In my experience, I have seen the positive effect of consistently applying brief mindfulness exercises with students others find difficult to teach. Let's look at the effect of mindfulness on one student and his family.

Wilson Applies School-Based Mindfulness at Home

Thirteen-year-old Wilson is sitting in his usual spot in the middle of the room, reading the morning newspaper at the beginning of the school day, while the diminutive art teacher is attempting to bring order to the fifty or so students and staff arranged around the large room. We are about to begin our daily mindful practice. Wilson, an obvious presence, newspaper open, is sitting directly across from the art teacher. She quietly and politely asks him to put the paper down so we can begin. He pointedly ignores her and noisily turns the page. She asks him again and he responds, "This mindful s—t is stupid. I can learn more from the newspaper." The art teacher has many years of experience and quickly realizes she has made a basic error, confronting a student in front of his peers where the student has to do nothing to win a power struggle. She recoups and announces that we will do our mindful moment without Wilson. Forging ahead, she explains that the mindful minute is a portable tool that students can use anywhere as long as they are breathing. "Close your eyes if you want or focus on the wall or the floor—you choose. Keep your focus on your breath—your mind will wander and Wilson will make noise with his newspaper—use any distraction as a way to practice returning your attention to your breath." The room becomes silent except for the newspaper as we all practice directing our attention to our breath, observing, labeling, and letting go of the emotions that arise whenever Wilson noisily turns another page. The mindfulness practice ends, a new school day begins, and the art teacher calmly says to Wilson, "Perhaps tomorrow?"

Later that week, Wilson sits reading his newspaper, but he turns the pages quietly. A few weeks later, I observe him sitting with his newspaper open on his lap, but his eyes are closed. A few months later in a coincidental meeting with his mother, she laughingly recalls: "Last night, my youngest and I got into another argument. I don't even remember what it was about now. Usually Wilson gets angry, too, and starts yelling 'shut up' from his room. Last night he charged in and said: 'You people need to take a mindful minute right now!' We were so surprised we stopped. I don't know who was more shocked, me or my youngest. Wilson yelled at us to breathe and count! He shouldn't yell at his mother, but you know, maybe it helped a little. Anyway what have you people done to my kid? Before he would have been angry all night. Now he is making us stop and do that mindful thing."

Wilson gradually became an advocate of mindfulness at school suggesting, usually loudly, that a class should take a mindful minute if it became too noisy. After his father left home, he confided in me that he used mindfulness to help calm his worries as he went to bed at night.

The Application of Mindfulness

Once Wilson discovered mindfulness could help him, he found his own applications at home. His heavy-handed application of the technique with his mother and brother can be excused for this socially awkward boy. Yet after introducing a few minutes of mindfulness repeated frequently in school almost a decade ago, I have found his story not all that unusual. Students find mindfulness personally useful and then spontaneously apply it at home, with sports, and before testing or ask parents to join them in mindful minutes. These student applications, crude as they may be, demonstrate their awareness of its significant benefits. Other applications are showing some initial scientific support. They are noted here because introducing a regular mindful practice into a classroom has many positive repercussions.

Mindfulness can moderate the effect of distressing events, decreasing the intensity of the emotions triggered by them and offering the possibility for a positive reappraisal. This appears possible because a consistent mindfulness practice asks us to repetitively notice thoughts, emotions, or sensations, observe them, and then let them go (Garland, Gaylord, and Park 2009). The repetitive practice trains the brain not to immediately become lost in thoughts, emotions, or sensations, but to step back, making it possible to shift our perspective and see them as a separate entity. By encouraging the lack of attachment to thoughts, emotions, and sensations, it liberates us to direct our awareness where we wish, opening the possibility of a positive reappraisal of events (Harris 2009). For example, in Wilson's case, his family's frequent arguing consistently triggered his anger. His school-based mindfulness practice enabled him to change his habitual response. Instead of becoming caught by a spiral of increasing anger, he was able to step back and change his response from anger to demanding his mother and brother use mindfulness to calm.

Wilson is just one example of what is possible for you and your students. A consistent personal mindfulness practice can enable you to change your habitual reaction to frustrating student misbehavior. It offers your students the opportunity of developing a state of mind that is conducive to learning and sound relationships.

Tools for School

1. Beginning Tomorrow. Beginning a mindfulness program can be as simple as starting your next class by asking your students to listen to silence for one minute. Time it, and if they are getting restless at thirty seconds, tell them time is up and lengthen it next time. Then ask them what they heard. Simple, short, and successful is the best way to start. Once it becomes a daily routine, add the instructions to follow your breath. After a minute or two, a quick talk about some of the benefits listed in this chapter can help as well. Later you may want to expand into guided meditations.

2. Begin with Your Colleagues. There is nothing better than a few minutes of silent mindfulness after a long day of teaching. A mindfulness practice with your colleagues can start for you the same way it started for me: just before the beginning of an after-school faculty meeting I impulsively said aloud: “It has been a rough day. Before we start the meeting, let’s just take a minute and listen to the silence. I’ll keep time.” I timed it for exactly one minute, and everybody appreciated the quiet after a day with noisy, busy kids. The principal was leading the meeting, and I probably should have asked him first, but the effect was so positive that it has been repeated ever since. More formal instruction on various methods followed this initial impulsive request.

3. Practice at Home. To effectively implement a mindfulness practice at school, it helps immeasurably to practice it at home. Mindfulness practice has many variations because of its rich history and current research. There are several websites that offer extensive resources:

- www.drdansiegel.com offers an interesting mindful meditation that combines neuroscience and mindfulness. Siegel calls it the Wheel of Awareness, and an audio version is available in the Resources section of his website.
- www.marc.ucla.edu. On the West Coast, the Mindful Awareness Research Center at UCLA is a resource for research results, a center for workshops and conferences, and a website with free audio-guided meditations from three minutes to twenty minutes.
- www.garrisoninstitute.org on the East Coast (New York) offers workshops and teacher training in mindfulness techniques.

- www.positivityresonance.com is the website associated with Barbara Fredrickson's book *Love 2.0* (2012) and offers several free audio-guided loving kindness meditations.
- www.susankaisergreenland.com. Susan Kaiser Greenland has been teaching mindfulness to young children for many years, incorporating brain research. Her website has video clips and other resources.

4. Other Resources. Kabat-Zinn's research results and MBSR techniques are accessible in his books. *Wherever You Go, There You Are: Mindfulness Meditation in Everyday Life* (1994) and *Coming to Our Senses: Healing Ourselves and the World through Mindfulness* (2005) are two of his classics. There are numerous other books about mindfulness. There are currently several thousand books on mindfulness available, with more being released each year.

Chapter 8

Putting It All Together: Classroom Culture

THERE IS A PARTIALLY TRUE ASSUMPTION MANY EDUCATORS HOLD: A TEACHers' primary job is to impart information to relatively passive student recipients. The value of the felt experience of school and classroom culture is minimized, even made invisible, by this assumption. The usual mode of thinking places the classroom teacher or the school principal at the center hub of a wheel directing students and teachers on the rim. But if you could actually see the connections in a classroom and school, they would look more like a web than a wheel with a central hub. The principal or headmaster of a school and the teacher in the classroom do not sit at the hub directing individuals; they are parts of an interactive web of relationships that include all members of the faculty and staff as well as students and parents. This invisible web of relationships creates school and classroom culture. Day to day, the culture of the organization is felt mostly through right hemisphere processes, which can make it difficult to discuss and seemingly lower its value to student learning, but it can be made visible, logically analyzed, and intentionally altered using left hemisphere processes. School and classroom culture is a collective invisible force that can be harnessed for the good of all its members.

No School Is an Island

School and classroom culture arises not only from the interaction of the minds, brains, and relationships within the school; it is influenced by the larger cultures of the city, state, and nation in which the school and classroom is embedded. For example, as is well known, contemporary state and national cultures are putting educators under considerable pressure to teach equally well to the gifted, the disabled, the anxious, the angry, and the non-native speakers, while neuroscience shows us that students arrive in our classrooms with immensely different patterns of neural connections already in place. The culture outside the school measures success using standardized tests that completely ignore the nonverbal unconscious lessons taught by the school and classroom culture as well as by the teacher. There are immense variations among school cultures nationwide, and they share the effect of the larger national culture. The pressure can be worrisome and, as you have learned, it can chronically activate your sympathetic nervous systems, narrowing your thinking.

Something seems to be missing from the important assessment of student achievement: the critically important yet invisible lessons learned from the school and classroom culture—the interpersonal, emotional, and character lessons taught within the school and classroom culture. Because so few educators understand how to address classroom culture directly, there are wide variations in cultures school to school, and there can be surprising variations among the cultures of different classrooms within the same school. I'm sure you have had the experience of walking through the halls of a school: some classrooms are quiet with nary a sound or movement; others are bustling with loud discussions and students in motion; in some the noise is a sign of disruption, in others it is a sign of engaged learning. The quiet of some classes is a sign of intense concentration; in others a sign of rigid rules inflexibly applied. They are the external signs of the internally felt experience of classroom culture.

Investigating school and classroom culture using what we have learned from previous chapters can bring it out of the realm of vaguely felt experience into a space where we can thoughtfully consider it and influence it for the betterment of students and ourselves. The bottom-up experience of a school or classroom culture may be simply defined as "the way things are around here" (Swift 2012). At the opposite end of the spectrum are the top-down attempts to define classroom culture: the Center for Social and Emotional Education's National School Climate Standards: with its five standards, fifteen indicators, and thirty subindicators (CSEE 2009), universal design for learning (CAST 2011), and to some extent Positive Behavioral Intervention and Supports (PBIS) (OSEP 2009) and Response to Intervention (RTI) (NCRTI, 2012).

Let's consider school and classroom culture not as an experientially true but vague bottom-up—"the way things are around here"—nor as a valuable but imposing set of top-down guidelines. Rather let's take the middle ground and consider classroom culture as something that includes the bottom-up lived life of the classroom yet is formal enough to offer useful top-down suggestions for analysis and application: "The social environment in which an individual lives. [It] shapes the context in which energy and information are shared among people by way of patterns of interactions, rituals of behavior, communicative symbols, and structural aspects of the environment" (Siegel 2012a, p. AI-21).

A school, a classroom, or for that matter any organization will inevitably develop its own culture, pattern of interactions, rituals of behavior, communicative symbols, and structural aspects so its members can share energy and information. In my experience, attending and altering these aspects of your school and classroom can create a positive culture for all of its members. If you have not considered that your classroom and school have their own cultures before, using Siegel's list ask yourself the following questions.

- *What is your classroom's pattern of interactions?* Do students raise their hands before speaking, or are there rollicking discussions? Do they talk with each other or just to you? What is your school's pattern of interactions among the faculty and administration? Do the members of the administration dictate policy? Are there rollicking discussions about a policy altering it before its adoption?
- *What are the rituals of behavior for your classes?* Do you stand in the hall and greet students as they arrive? Take attendance at the beginning of each class? Begin with a few mindful minutes? What are the rituals of behavior in your school among the faculty and administration? Are there mandatory scheduled interminable faculty meetings whether or not there is anything to discuss? Are meetings scheduled only when there is a crisis? Is it something in between? Do teachers gather at a local watering hole regularly?
- *What are the communicative symbols?* Are your classroom walls covered with completed work and posters, or bare and free of visual distraction? Does your school have signs and a website with the mission statement prominently displayed? Is there a bulletin board listing students' accomplishments? Showing athletic accomplishments? Faculty accomplishments?
- *What about structural aspects?* Are the desks in rows or the chairs in circles around tables? Is your school building old

and pridefully historic or old and worn out? Is the building well designed for the size of enrollment? If there is a piece of trash in the parking lot will you, another teacher, or student be likely to pick it up?

It may appear to be a paradox that attending and influencing something as broad as classroom or school culture will enhance individual student learning, but there is an expanding body of evidence that supports the concept (Cohen 2006; National School Climate Council, 2007; Cohen et al. 2009). It is also not a new idea. The National School Climate Council dates the beginning of interest in classroom climate to a book published in 1908: *The Management of a City School* by A. Perry.

Real-Life Example of School and Classroom Culture

It is worth describing an example of the development of a positive school culture with all the quirks that a real-world application entails. Seen in hindsight, the turning point for creating a positive, strengths-based school culture in one school offers a peek into how all the moving parts of cultural development actually work.

The turning point occurred when this school's culture expanded from a partially effective but ultimately negative focus on eliminating weaknesses (such as behavioral, emotional, and learning challenges) to focus on increasing the positive by recognizing and expanding strengths. This began several years ago in a faculty meeting where we were yet again trying to sort out the myriad problems that arise when you teach the most difficult students. It began in the depths of a cold, dark winter; as I looked around the room, I saw tired teachers trying to solve difficult and maybe unsolvable problems with hard-to-teach students. As we discussed Liam's lack of progress in math, his annoying arrogance, and lackluster performance in other classes, I became desperate: "He must be good at something. What is Liam good at?" The director, probably feeling as desperate as I did, interrupted the meeting and pointedly asked each teacher to take a turn and describe one of Liam's strengths or something positive about him. Half-heartedly, each teacher took a turn. Someone mentioned his love of reading, someone else noted that his peers liked him. With a laugh, we realized how difficult it was to change our point of view when someone mentioned he had a nice hairstyle. As we went on to discuss other students, we forced ourselves to talk about their strengths and interests before the current problems they presented. The bleak mood in the room slowly shifted, and laughter could be heard.

Remembering a well-researched activity introduced in the positive psychology class I had just completed, I suggested we have a school-wide writing assignment called "the me at my best story." The directions for the assignment are deceptively simple: "write a one-page story that describes a specific event in your life when you felt you were at your best. The story should have a beginning, a middle, and end with a bang" (Seligman 2002). I said what at the time seemed like an off-hand remark: "We would never do anything to our kids that we would not do to ourselves first, right?" The director told everyone to come to the next week's staff meeting with a "me at my best" story.

The next week the director read his story first, I read mine, and a few teachers followed, but because of time constraints we moved onto other, more pressing business. Each staff meeting for many weeks began with a few teachers reading their stories, which enabled us to listen intently, generating a prolonged and increased positive effect of the stories on the staff culture. (As expected in a novel experience like this one, not everyone read theirs and not everyone did the assignment.) We drew closer together and developed increased trust as we learned about the high points in each other's life.

The faculty's positive experience with writing and reading the stories meant that they introduced the assignment to the students with enthusiasm. Some appropriately used their own stories as examples for the students. Some students wrote moving stories they willingly read in class, for some it was an adventure in self-discovery, for a few it was as expected just another assignment to complain about.

The assignment began the change toward focusing on the positive. In actuality the positive was always there, as it is in any school, and this exercise shined a spotlight on it and brought positivity to life. Nearly a decade later the school continues to intentionally develop and maintain a positive school and classroom culture by beginning most staff meetings with positive comments, daily mindfulness, a focus on student and teacher strengths, and other means.

It can at times seem that the larger national culture is continuously bombarding us with negative information. As discussed earlier, older neurological systems continually scan the environment for potential threats. These older systems do not "know" whether the potential threats are real and present or just images on the television or figments of the imagination. Purposely emphasizing safety, connection, and positivity at school can help create a cultural island where students and teachers can learn and flourish.

School Culture Fosters Neural Integration

Dan Siegel defines neural integration as the linkage of differentiated parts, each part a separate unique entity functioning at its best with solid linkages to the other well-functioning parts (Siegel 2012b). Integration can be considered an index of health and well-being. A healthy brain has well-functioning parts well linked to its other parts. A single damaged or poorly developed brain region (resulting from brain damage or trauma during childhood, for example) can impede the whole brain from functioning at an optimal level. A damaged or poorly developed link between brain regions, even if those parts function well on their own, also stops the brain from functioning well.

The benefits of integration—highly functioning parts communicating well with each other—can be seen outside of the internal workings of the brain, for example, in a family. "Secure attachment involves both the differentiation of child from parent and the empathic and attuned communication between the two. Suboptimal attachment, as we've seen, involves impediments to this differentiation and linkage" (Siegel 2012a, p. 354). Like a well integrated brain the highly functioning parts of a family (children and parents) combined with an empathic connection between them leads to well-being for all members. It is similar in a healthy classroom: individual students learn well and are empathetically linked to others in the class, including you. Educators know all too well the negative effect that a single poorly functioning student can have on a whole class (an example of lack of integration). You may have experienced another example of the lack of integration, a class of highly functioning individual students who were antagonistic to each other and are draining to teach. Their separate, individual skills do not make up for the problematic linkages among them. I sincerely hope you have had the experience of a class full of high-functioning students who also work well together—this combination makes teaching a joy. The easy flow of teaching such a class is like being carried by the current of a river.

The River of Classroom and Neural Integration

Siegel describes integration using the metaphor of a river for good reason. His river of integration flows between the two banks of chaos and rigidity. The chaos is caused by differentiated parts with no connection, and the rigidity is caused by connection that ignores individual differences. You probably have seen classes with teaching that is closer to one bank than the other: the chaotic noisy classroom where the teacher is being pulled in multiple directions simultaneously by each student and does not address the class as a whole, or the overly quiet, rigid classroom where the same lessons are monotonously taught the same way using the same worksheets and questions year after year without regard to student differences.

One day a principal I know taught me the difference between these two extremes. I was helping hire a new science teacher and was excited about one candidate's ten years of experience in the same school, which to me meant she probably could add depth of experience to the current faculty. He abruptly calmed my enthusiasm and described the bank of rigidity in a single sentence when he said, "A teacher can have ten years of valuable experience, or they can have one year of experience repeated ten times. There is a big difference." In other words, was her teaching experience within the center of the river of integration, or was it along the banks of rigidity and repetition?

The truth for excellent, experienced teachers is that their teaching during times of creative discussion probably drifts toward the bank of chaos and during times when students must be taught something that requires repetition and drill may drift toward the bank of rigidity. A healthy classroom that fosters integration among the students intentionally flows in a winding course.

Keeping the metaphor of the river of integration in mind can help create a state of mind for educators from which effective interventions and techniques arise. Keeping a classroom and school within the flow near the center of the river is critically important. A classroom and school that flows near the center of the river of integration supports the well-being of all its members. Students and teachers feel safe when they are far from the riverbank of chaos caused by disruptive misbehavior, bullying, and violence. They are open to new learning when they are far from the riverbank of rigidity caused by overly strict rules, inflexible policies, and repetitive, uninspiring lessons. A classroom and school that is within the flow of integration fosters the essential human connections described in Chapter 3.

Secure attachment is only possible far from the bank of chaos, with its fear-producing unpredictability, and far from the bank of rigidity, with its harsh and sometimes abusive rule enforcement. A classroom and school in the flow of integration helps students focus their attention on learning. One that drifts toward the bank of chaos makes it extremely difficult for students to control their attention because it is continually being grabbed by the unpredictability that defines chaos.

Rigid, dull repetitive experiences (near the bank of rigidity) make it equally difficult for students and teachers to care enough to invest the energy necessary to direct their attention to learning. The current near the center of the river maintains the balance necessary for students to have positive experiences in school that will be stored in implicit and explicit memory. Chaos can create implicit memories that trigger anxiety and fear even at the thought of attending school. Rigidity and repetition can make it difficult to remember the difference from one day to another, similar to the example of your commute to school used in Chapter 5.

You can maintain your classroom and school within the flow at the center of the river by focusing on student and teacher strengths. Intentionally helping students and teachers discover and expand their strengths as well as improve their weaknesses keeps everyone away from the banks of chaos and rigidity. Adding a strength focus creates a balance to the built-in weakness improvement approach in education. It is neither the chaos caused by ignoring either strengths or weaknesses, nor is it a rigid adherence to only weakness improvement or only strength development—it is a combination of both.

Finally, mindfulness is a personal practice that can be the keel that keeps your boat steady in the center of the river. Spending time focusing on your breath while you label and let go of thoughts and emotions when they inevitably arise in your mind—this practice keeps you away from both banks. It does the same for your students.

Changing Classroom and School Culture

Classroom and school culture behave like a big cruise ship. Ocean liners easily set their course while getting under way and leaving port but can only change course slowly once they are steaming full speed ahead. Like the cruise ship, the course of the classroom culture is best set at the beginning of the school year or some other natural break, such as a new semester or after a vacation. At the beginning of the school year, less energy is required to set course, but a course change is possible at any time. Most teachers begin to create a classroom culture by establishing rules, expectations, and consequences up front. They are one important aspect of the classroom culture, but a lot of emphasis on them will pull the culture close to the bank of rigidity. Balancing this necessary task by establishing a routine of a few minutes of mindfulness (see Chapter 7) and a strengths exercise (see Chapter 6) will help your classroom culture set a middle-of-the-river course.

To begin to alter course any time during the year, try beginning with a few minutes of mindfulness. You probably have noted throughout this book that classroom and school cultural change begins with the adults and is then followed by the students. To add a strength focus, teachers and administrators need to know their own strengths. To use mindfulness as a practice in the classroom, teachers and administrators need to practice it themselves.

Begin with yourself, followed by your colleagues and administrators, and end with your students.

References

Aimone, J. B., Wiles, J., and Gage, F. H. (2006). Potential Role for Adult Neurogenesis in the Encoding of Time in New Memories. *Nature Neuroscience, 9*(6), 723–27.

Ainsworth, M. D. S., Blehar, J. C., Waters, E. and Wall, S. (1978). *Patterns of Attachment: A Psychological Study of the Strange Situation.* Hillsdale, NJ: Erlbaum.

Armstrong, K. (1993). *The History of God: The 4000 Year Quest of Judaism, Christianity and Islam.* New York: Random House.

Badenoch, B. (2008). *Being a Brain-Wise Therapist: A Practical Guide to Interpersonal Neurobiology.* New York: W. W. Norton and Company.

Badenoch, B. (2011). *The Brain-Savvy Therapist Workbook: A Companion to Being a Brain-Wise Therapist.* New York: W. W. Norton and Company.

Badenoch, B. (2013). Refining Our Understanding of the Hemispheres. Webinar, November 14, 2013. Transcript available at http://mindgains.org/index.php.

Baumeister, R. F., Bratslavsky, E., Finkenauer, C., and Vohs, K. D. (2001). Bad Is Stronger than Good. *Review of General Psychology, 5*(4), 323–70.

Baumeister, R. F. and Tierney J. (2011). *Willpower: Rediscovering the Greatest Human Strength.* New York: Penguin.

Beauchemin, J., Hutchins T. L., and Patterson L. (2008). Mindfulness Meditation May Lessen Anxiety, Promote Social Skills, and Improve Academic Performance Among Adolescents with Learning Disabilities. *Complementary Health Practice Review*, 13, 34-45.

Beckes, L., and Coan, J. A. (2011). Social Baseline Theory: The Role of Social Proximity in Emotion and Economy of Action. *Social and Personality Psychology Compass,* 5(12), 976–88.

Bergin, C., and Bergin, D. (2009). Attachment in the Classroom. *Educational Psychology Review*, 21, 141–170.

Bisley, J. W. (2011). The Neural Basis of Visual Attention. *Journal of Physiology*, 589, 49–57.

Black, D. S., Milan, J., and Sussman, S. (2009). Sitting-Meditation Interventions among Youth: A Review of Treatment Efficacy. *Pediatrics,* 124(3), 532–41.

Blumenfeld, P. C., Soloway, R. W., Krajcik, M. G., and Palincsar, A. (1991). Motivating Project-Based Learning: Sustaining the Doing, Supporting the Learning. *Educational Psychologist,* 26(3 and 4), 369–98.

Bowlby, J. (1982). *Attachment and Loss: Vol. 1. Attachment* (2nd ed.). New York: Basic Books.

Bowlby, J. (1973). *Attachment and Loss: Vol. 2. Separation*. New York: Basic Books.

Bowlby, J. (1980). *Attachment and Loss: Vol. 3. Loss, Sadness and Depression*. New York: Basic Books.

Brown, B. (2010). *The Gifts of Imperfection: Let Go of Who You Think You're Supposed to Be and Embrace Who You Are*. Minnesota: Hazelden.

Brown, B. (2012). *Daring Greatly: How the Courage to Be Vulnerable Transforms the Way We Live, Love, Parent, and Lead*. New York: Gotham Books.

Bryk, A. S., and Schneider, B. (2002). *Trust in Schools: A Core Resource for Improvement*. New York: Russell Sage Foundation.

Buckingham, M., and Clifton, D. O. (2001). *Now, Discover Your Strengths*. New York: Free Press.

Buron, K. D., and Curtis, M. (2012). *Incredible 5 Point Scale: The Significantly Improved and Expanded Second Edition; Assisting Students in Understanding Social Interactions and Controlling their Emotional Responses*. Shawneee Mission, KS: AAPC Publishing.

Buschman, T. J., and Miller, E. K. (2007). Top-Down versus Bottom-Up Control of Attention in the Prefrontal and Posterior Parietal Cortices. *Science,* 315(5820), 1860–62.

CAST. (2011). *Universal Design for Learning Guidelines version 2.0.* Wakefield, MA: CAST.

Coan, J. A. (2010). Adult Attachment and the Brain. *Journal of Social and Personal Relationships,* 27, 210–18.

Coan, J. A., Schaefer, H. S., and Davidson, R. J. (2006). Lending a Hand: Social Regulation of the Neural Response to Threat. *Psychological Science,* 17, 1032–39.

Cohen, J. (2006). Social, Emotional, Ethical and Academic Education: Creating a Climate for Learning, Participation in Democracy and Well-Being. *Harvard Educational Review,* 76, 201–37.

Cohen, J., McCabe, L., Michelli, N. M., and Pickeral, T. (2009). School Climate: Research, Policy, Practice, and Teacher Education. *Teachers College Record*, 111, 180–213.

Cooperrider, D., and Whitney, D. (2005). *Appreciative Inquiry: A Positive Revolution in Change*. San Francisco: Berrett-Koehler.

Cozolino, L. (2013). *The Social Neuroscience of Education: Optimizing Attachment and Learning in the Classroom*. New York: W. W. Norton and Company.

Creswell, D. J., Irwin, M. R., Burklund, L. J., Lieberman, M. D., Arevalo, J. M. G., Ma, J., Breen, E. C., and Cole, S. W. (2012). Mindfulness-Based Stress Reduction Training Reduces Loneliness and Pro-Inflammatory Gene Expression in Older Adults: A Small Randomized Controlled Trial. *Brain, Behavior, and Immunity*, 26, 1095–101.

Csikszentmihalyi, M. (1990). *Flow: The Psychology of Optimal Experience*. New York: HarperCollins.

Csikszentmihalyi, M and Nakamura, J (2010). Effortless Attention in Everyday Life: A Systematic Phenomenology. In Bruya, B. (Ed.), *Effortless Attention: A New Perspective in the Cognitive Science of Attention and Action* (179-189). Cambridge, MA: MIT Press.

CSEE. (2009). National School Climate Standards: Benchmarks to Promote Effective Teaching, Learning and Comprehensive School Improvement. Center for Social and Emotional Education, www.schoolclimate.org (retrieved July 2012).

Davidson, R. J., Kabat-Zinn, J., Schumacher, J., Rosenkranz, M., Muller, D., Santorelli, S. F., Urbanowski, F., Harrington, A., Bonus, K., and Sheridan, J. F. (2003). Alterations in Brain and Immune Function Produced by Mindfulness Meditation. *Psychosomatic Medicine*, 65(4), 564–70.

Davidson, R. J., and Lutz, A. (2008). Buddha's Brain: Neuroplasticity and Meditation. *IEEE Signal Process Magazine* 25(1), 174–76.

DeNisco, A. (2013). Homework or Not? That Is the (Research) Question. *District Administrator*, March, http://www.districtadministration.com/article/homework-or-not-research-question (retrieved November 2013).

Doidge, N (2007). *The Brain That Changes Itself*. New York: Penguin.

Drucker, P. F. (1967, 2006). *The Effective Executive: The Definitive Guide to Getting the Right Things Done*. New York: HarperCollins.

DuFour, R. (2004). Schools as Learning Communities. *Educational Leadership*, 61(8), 6–11.

Eades, J. M. F. (2008). *Celebrating Strengths: Building Strength-Based Schools*. Coventry, UK: CAPP Press.

Ecker, B., Ticic, R., Hulley, L., and Niemeyer, R. A. (2012). *Unlocking the Emotional Brain: Eliminating Symptoms at Their Roots Using Memory Reconsolidation*. New York: Routledge, Taylor and Francis.

Eisenberger, N. I., Taylor, S. E., Gable, S. L., Hilmert, C. J., and Lieberman, M. D. (2007). Neural Pathways Link Social Support to Attenuated Neuroendocrine Stress Responses. *NeuroImage*, 35, 1601–2.

Field, T., Diego, M., and Hernandez-Reif, M. (2006). Prenatal Depression Effects on the Fetus and Newborn: A Review. *Infant Behavior and Development*, 29(3), 445–55.

Fraley, R. C., Waller, N. G., and Brennan, K. A. (2000). An Item Response Theory of Self-Report Measures of Adult Attachment. *Journal of Personality and Social Psychology*, 78(2), 350–65.

Fredrickson, B. (2009). *Positivity: Groundbreaking Research Reveals How to Embrace the Hidden Strength of Positive Emotions, Overcome Negativity, and Thrive*. New York: Crown.

Fredrickson, B. (2012). *Love 2.0: How Our Supreme Emotion Affects Everything We Feel, Think, Do, and Become*. New York: Hudson Street Press.

Gallup. (n.d.). Education: Giving Education Leaders Tools and Advice to Help Teachers, Students, and Schools Succeed. Retrieved from http://www.gallup.com/strategicconsulting/en-us/education.aspx.

Garland, E. L., and Fredrickson, B. L. (2013). Mindfulness Broadens Awareness and Builds Meaning at the Attention-Emotion Interface. In T. B. Kashdan and J. Ciarrochi, eds., *Mindfulness, Acceptance and Positive Psychology*. Oakland, CA: Context Press.

Garland, E., Gaylord, S., and Park, J. (2009). The Role of Mindfulness in Positive Reappraisal. *Explore (NY)*, 5(1), 37–44.

Goleman, D. (2011). *The Brain and Emotional Intelligence*. Northampton, MA: More Than Sound.

Gottman, J. M., and Silver, N. (1999). *The Seven Principles for Making Marriage Work*. New York: Three Rivers Press.

Greenleaf, R. K., and Spears, L. C. (eds.). (1998). *The Power of Servant Leadership*. San Francisco: Barrett-Koehler.

Grabenhorst, F., and Rolls, E. T. (2008). Selective Attention to Affective Value Alters How the Brain Processes Taste Stimuli. *European Journal of Neuroscience*, 27(3), 723–29.

Gregory, A., and Weinstein, R. S. (2004). Connection and Regulation at Home and in School: Predicting Growth in Achievement for Adolescents. *Journal of Adolescent Research*, 19, 405–27.

Harris, R. (2009). *ACT Made Simple*. Oakland, CA: New Harbinger Press.

Hasson, U., Ghazanfar, A. A., Galantucci, B., Garrod, S., and Keysers, C. (2011) Brain-to-Brain Coupling: A Mechanism for Creating and Sharing a Social World. *Trends in Cognitive Science*, 16(2): 114–21.

Hebb, D. (1949). *The Organization of Behavior: A Neuropsychological Theory*. New York: Wiley.

Hesse, E. (2008) The Adult Attachment Interview: Historical and Current Perspectives. In Cassidy, J. and Shaver, P. R. (eds.), *Attachment Theory, Research and Clinical Applications* (2nd ed.). New York: Guilford Press.

Hodges, T. (2012). The New Definition of Career Readiness. Retrieved from http://thegallupblog.gallup.com/2012/10/the-new-definition-of-career-readiness.html.

Hord, S. A. and Sommers, W. A. (2008). *Leading Professional Learning Communities: Voices from Research and Practice*. Thousand Oaks, CA: SAGE.

Iacoboni, M. (2008). *Mirroring People: The Science of Empathy and How We Connect with Others*. New York: Farrar, Straus and Giroux.

Iacoboni, M. (2009). Imitation, Empathy, and Mirror Neurons. *Annual Review of Psychology*, 60, 653–70.

Jha, A. P. (2013). Being in the Now. *Scientific American Mind*, 24(1).

Jha, A. P., Krompinger, J., and Baime, M. A. (2007). Mindfulness Training Modifies Subsystems of Attention. *Cognitive Affective and Behavioral Neuroscience*, 7(2), 109–19.

Jha, A. P., Stanley, E. A., Kiyonaga, A., Wong, L., and Gelfand, L. (2010). Examining the Protective Effects of Mindfulness Training on Working Memory Capacity and Affective Experience. *Emotion*, 10(1), 54–64.

Jing, J., Dai, B., Peng, D., Zhu, C., Liu, L., and Lu, C. (2012). Neural Synchronization during Face-to-Face Communication. *Journal of Neuroscience*, 32 (45), 16064–69.

Kabat-Zinn, J. (1982). An Outpatient Program in Behavioral Medicine for Chronic Pain Patients Based on the Practice of Mindfulness Meditation: Theoretical Considerations and Preliminary Results. *General Hospital Psychiatry*, 4 (1), 33–47.

Kabat-Zinn, J. (1994). *Wherever You Go, There You Are: Mindfulness Meditation in Everyday Life*. New York: Hyperion.

Kabat-Zinn, J. (2005). *Coming to Our Senses: Healing Ourselves and the World Through Mindfulness*. New York: Hyperion.

Kagan, J. (1994). *Galen's Prophecy: Temperament in Human Nature*. Boulder, CO: Westview Press.

Kintsch, W., and Bates, E. (1977). Recognition Memory for Statements from a College Lecture. Institute for the Study of Intellectual Behavior, Memory Report 53. University of Colorado.

Klem, A. M., and Connell, J. P. (2004). Relationships Matter: Linking Teacher Support to Student Engagement and Achievement. *Journal of School Health*, 74(7).

Kohn, A. (2006). *The Homework Myth: Why Kids Get Too Much of a Bad Thing*. Philadelphia: Da Capo Press.

Kuypers, L. (2011). *The Zones of Regulation: A Curriculum Designed to Foster Self Regulation and Emotional Control*. San Jose, CA: Social Thinking Press.

Lazar, S. W., Kerr, C. E. Wasserman, R. H., Gray, J. R., Greve, D. N., Treadway, M. T., McGarvey, M., Quinn, B. T., Dusek, J. T., Benson, H., Rauch, S. L., Moore, C. I., and Fischl, D. (2005). Meditation Experience Is Associated with Increased Cortical Thickness. *Neuroreport*, 16(17), 1893–97.

Lieberman, M. (2013). *Social: Why Our Brains Are Wired to Connect*. New York: Crown.

Linley, A. (2008). *Average to A+: Realising Strengths in Yourself and Others*. Coventry, UK: CAPP Press.

Linley, A., Willars, J., and Biswas-Diener, R. (2010). *The Strengths Book*. Coventry, UK: CAPP Press.

Lopez, S. (2013). Hope, Academic Success, and the Gallup Student Poll. Retrieved from http://www.gallupstudentpoll.com/122168/Hope-Academic-Success-Gallup-Student-Poll.aspx.

Main, M., and Hesse, E. (1990). Parents' Unresolved Traumatic Experiences Are Related to Infant Disorganized Attachment Status: Is Frightened and/or Frightening Parental Behavior the Linking Mechanism? In Greenberg, M. T., Cicchetti, D., and Cummings, E. M. (Eds.). *Attachment in the Preschool Years: Theory, Research, and Intervention.* Chicago: University of Chicago Press.

Main, M., Hesse, E., and Goldwyn, R. (2008). Studying Difference in Language Usage in Recounting Attachment History: An Introduction to the AAI. In Steele, H., and Steele, M. (eds.), *Clinical applications of Adult Attachment Interview* (pp. 31–68). New York: Guilford Press.

Maslow, A (1954). *Motivation and Personality.* New York: Harper.

McGilchrist, I. (2009). *The Master and His Emissary: The Divided Brain and the Making of the Western World.* New Haven, CT: Yale University Press.

Meichenbaum, D. H., and Goodman, J. (1971). Training Impulsive Children to Talk to Themselves: A Means of Developing Self-Control. *Journal of Abnormal Psychology*, 77(2), 115–26.

Napolia, M., Krech, P. R., and Holley, L. C. (2005). Mindfulness Training for Elementary School Students: The Attention Academy. *Journal of Applied School Psychology*, 21 (1).

National School Climate Council. (2007). The School Climate Challenge: Narrowing the Gap between School Climate Research and School Climate Policy, Practice Guidelines and Teacher Education Policy. Retrieved from www.schoolclimate.org/climate/policy.php/.

NCRTI. (2012). What Is RTI? National Center on Response to Intervention. http://www.rti4success.org/categorycontents/multi-level_prevention_system (retrieved July 2012).

Nummenmaa, L., Glerean, E., Viinikainenb, M., Jääskeläinenb,I. P., Haria, R. and Sams, M. (2012) Emotions promote social interaction by synchronizing brain activity across individuals. *Proceedings of the National Academy of Science.* 109(24), 9599-9604.

Olson, K. (2006). Contingency in the Classroom. *Connections and Reflections: The GAINS Quarterly*, part II, 1–3.

OSEP. (2009). What Is School Wide Positive Behavioral Intervention and Support? Office of Special Education Programs: Center on Positive Behavioral Interventions and Supports. http://www.pbis.org/common/cms/documents/WhatIsPBIS/WhatIsSWPBS.pdf (retrieved July 2012).

Palmer, P. J. (2007). *The Courage to Teach: Exploring the Inner Landscape of a Teacher's Life*, 10th anniversary edition. San Francisco: Jossey-Bass.

Palmer, P. J. (2004). *A Hidden Wholeness: The Journey Toward and Undivided Life*. San Francisco: Jossey-Bass.

Panksepp, J., and Biven, L. (2012). *The Archaeology of Mind: Neuroevolutionary Origins of Human Emotion*. New York: W. W. Norton and Company.

Peterson, C. (2006). *A Primer in Positive Psychology*. New York: Oxford University Press.

Peterson, C., and Seligman, E. (2004). *Character Strengths and Virtues: A Handbook and Classification*. New York: Oxford University Press.

Porges, S. W. (2011). *The Polyvagal Theory: Neurophysiological Foundations of Emotions, Attachment, Communication, and Self-Regulation*. New York: W. W. Norton and Company.

Ratey, J. J., and Hagerman, E. (2008). *Spark: The Revolutionary New Science of Exercise and the Brain*. New York: Little Brown.

Rath, T. (2007). *Strengthsfinder 2.0*. New York: Gallup Press.

Rath, T., and Harter, J. (2010). *Well Being: The Five Essential Elements*. New York: Gallup Press.

Reio, T. G., Marcus, R. F., and Sanders-Reio, J (2009). Contribution of Student and Instructor Relationships and Attachment Style to School Completion. *Journal of Genetic Psychology*, 170(1), 53–72.

Riley, P. (2011). *Attachment Theory and the Teacher-Student Relationship: A Practical Guide for Teacher Educators and School Leaders*. London: Routledge.

Rosenthal, R., and Jacobs, L. (1968). Pygmalion in the Classroom. *Urban Review*, 3(1), 16–20.

Schnall, S., Harber, K., Stefanucci, J., and Proffitt, D. R. (2008). Social Support and the Perception of Geographical Slant. *Journal of Experimental Social Psychology*, 44, 1246–55.

Schore, A. N. (2012). *The Science of the Art of Psychotherapy*. New York: W. W. Norton and Company.

Schwartz, J., and Begley, S. (2003). *The Mind and the Brain: Neuroplasticity and the Power of Mental Force*. New York: Regan Books.

Seligman, M. E. P. (1990). *Learned Optimism: How to Change Your Mind and Your Life*. New York: Vintage Books.

Seligman, M. E. P. (1995). *The Optimistic Child: A Proven Program to Safeguard Children against Depression and Build Lifelong Resilience*. New York: Houghton Mifflin.

Seligman, M. E. P. (2002). *Authentic Happiness: Using the New Positive Psychology to Realize Your Potential for Lasting Fulfillment*. New York: Free Press.

Seligman, M. E. P. (2011). *Flourish: A Visionary New Understanding of Happiness and Well-being*. New York: Simon and Schuster.

Shatz, C. J. (1992). The Developing Brain. *Scientific American*, 267(3), 60–67.

Siegel, D. J. (1999). *The Developing Mind: How Relationships and the Brain Interact to Shape Who We Are*. New York: Guilford Press.

Siegel, D. J. (2007). *The Mindful Brain: Reflection and Attunement in the Cultivation of Well-Being*. New York: W. W. Norton and Company.

Siegel, D. J. (2010a). *Mindsight: The New Science of Personal Transformation*. New York: Random House.

Siegel, D. J. (2010b). *The Mindful Therapist: A Clinician's Guide to Mindsight and Neural Integration*. New York: W. W. Norton and Company.

Siegel, D. J. (2012a). *The Developing Mind: How Relationships and the Brain Interact to Shape Who We Are* (2nd ed.). New York: Guilford Press.

Siegel, D. J. (2012b). *Pocket Guide to Interpersonal Neurobiology*. New York: W. W. Norton and Company.

Siegel, D. J. (2013). *Brainstorm: The Power and the Purpose of the Teenage Brain*. New York: Penguin.

Siegel, R. D. (2010). *The Mindfulness Solution: Everyday Practices for Everyday Solutions*. New York: Guilford Press.

Sroufe, S., and Siegel, D. J. (2011). The Verdict Is In: The Case for Attachment Theory. *Psychotherapy Networker*, March/April.

StrengthQuest (2000). Reference Card. Gallup Press. Retrieved from www.Strengthsquest.com.

Stephens, G. J. , Silbert, L. J., and Hasson, U. (2010). Speaker – listener Neural Coupling Underlies Successful Communication. *Proceedings of the National Academy of Science*, 10(32), 14425-14430.

Swift, S. (2012). An ELT Notebook: First Lessons Establishing Classroom Culture. Retrieved from http://eltnotebook.blogspot.com.

Tronick, E. (2007). *The Neurobehavioral and Social-Emotional Development of Infants and Children*. New York: W. W. Norton and Company.

Van Praag, H., Kempermann, G., and Gage, F. (1999). Running Increases Cell Proliferation and Neurogenesis in the Adult Mouse Dentate Gyrus. *Nature Neuroscience*, 2(3), 266–70.

Wang, J. S., Knipling, R. R., Goodman, M. J. (1996). The Role of Driver Inattention in Crashes: New Statistics from the 1995 Crashworthiness Data System. *40th Annual Proceedings of Association for the Advancement of Automotive Medicine*, October 7–9, Vancouver, BC.

Wisner, B. L., Jones, B., and Gwin, D. (2010). School Based Meditation Practices for Adolescents: A Resource for Strengthening Self-Regulation, Emotional Coping, and Self-Esteem. *Children and Schools*, 32(3).

Yeager, J. M., Fisher, S. W., and Shearon, D. N. (2011). *SMART Strengths: Building Character, Resilience and Relationships in Youth*. Putnam Valley, NY: Kravis.

Yun, K. (2013). On the Same Wavelength: Face-to-Face Communication Increases Interpersonal Neural Synchronization. *Journal of Neuroscience*, 33 (12), 5081–82.

Index

In this index, *f* denotes figure.